Photo Guide to Birds of Costa Rica

Photo Guide to Birds of Costa Rica

RICHARD GARRIGUES

A Zona Tropical Publication

FROM

Comstock Publishing Associates

a division of

Cornell University Press

Ithaca and London

First published 2015 by Cornell Universtity Press

First printing, Cornell Paperbacks, 2015
Printed in China

Library of Congress Cataloging-in-Publication Data

Garrigues, Richard, author.
Photo guide to birds of Costa Rica: / Richard Garrigues.
pages cm
Includes index.
ISBN 978-1-5017-0025-5 (pbk. : alk. paper)
1. Birds—Costa Rica—Pictorial works. I. Title.
QL687.C8.G38 2015
598.097286–dc23

2015015122

Zona Tropical ISBN 978-0-9816028-9-9

Cornell University Press strives to use environmentally responsible suppliers and materials to the fullest extent possible in the publishing of its books. Such materials include vegetable-based, low-VOC inks and acid-free papers that are recycled, totally chlorine-free, or partly composed of nonwood fibers. For further information, visit our website at www.cornellpress.cornell.edu.

Paperback printing 10 9 8 7 6 5 4 3 2 1

Book design: Gabriela Wattson

This book is dedicated to the memory of Alexander F. Skutch, who learned so much about the lives of birds through simple, patient observation and shared that knowledge via his wonderful prose.

Contents

Introduction

This book is designed for two distinct sets of readers. First, are birders new to birding—or new to birding in Costa Rica—who want a guide to the birds that one is most likely to see, as well as to a few of the rarer species that one would hope to encounter. It treats over 40% (365) of the species known from CR, but is a guide to at least 75% of the birds commonly seen in a week or so of birding—and an even higher percentage if birding is not your primary objective. Second, are experienced birders in search of a companion volume to the second edition of *The Birds of Costa Rica* (Garrigues and Dean), an illustrated guide to all the birds of Costa Rica. Birders, it seems, are hungry for information, and the excellent photos (549 in total) in this book will prove to be a valuable supplement to the illustrations in Garrigues and Dean.

Biologists categorize all organisms by placing them in related groups. Without going into unnecessary detail, suffice it to say that birds are divided into families, and then further defined to the level of genus and species—just as humans (genus: *Homo*, species: *sapiens*) are members of the primate family. Each group of birds in this book is introduced by a brief description of the principal characteristics of the family; at the end of each family description is the current number of species worldwide, followed by the number of species found in Costa Rica (i.e., World: 48, CR: 20). These numbers are based on current taxonomy, but family revisions by taxonomists can cause the numbers to change.

To be in accordance with eBird—a powerful and important online data-collection tool—we chose to follow the nomenclature and systematics of the eBird/Clements Checklist of Birds of the World: Version 6.9. However, to make this book more user-friendly for those who are not already familiar with bird taxonomy, we have re-grouped families so that birds of somewhat similar appearance and behavior are in close proximity (even if we are told that they are not at all closely related evolutionarily).

The order of families in roughly the first third of the book is as follows: game birds, swimming birds, aerial pelagic and coastal species, wading birds and shorebirds, diurnal birds of prey, nocturnal birds, aerial insectivores, and hummingbirds. From there, the families follow more closely the order prescribed by the eBird/Clements list.

Following the family description are the species accounts, which contain the following elements:

Species name. Each species is identified by its English common name and scientific name (in italics), according to the eBird/Clements Checklist of Birds of the World: Version 6.9. Both names appear in the Species Index (p. 245).

Measurements. Each account begins with a measurement of the length of the bird from the tip of the bill to the tip of the tail. Where there is a notable difference in size between the sexes, we give the length of both. The measurements have been rounded off to the nearest whole figure and should only be used as a general indication of size. Judging the size of a bird can be extremely challenging, and it's important to remember that many factors influence the impression of a bird's size—lighting, distance from the observer, and whether or not it is viewed through optical equipment. Similarly, while the length measurements indicate size, they do not account for leg length or a bird's shape and mass—all factors that can definitely affect the impression of size.

Two 5-inch birds show that length measurements alone can be a misleading indication of size.

Field marks. After the measurements, we describe pertinent field marks that distinguish each species. Sometimes mention is made of a similar species in the same genus that is not illustrated, in which case the common name is given, followed by the first letter of the genus and the full specific epithet (e.g., Plain Chachalaca *O. vetula*). Please consult the Anatomical Features section (p. 8) to clarify any body-part terms that may not be familiar to you. For definitions of non-anatomical terms, see the glossary on p. 14.

Habitat and behavior. This will give you an idea of where to expect to find a species within its mapped range, as well as offer an insight into how the bird might behave (e.g. foraging at a certain level of the forest, accompanying mixed species flocks, or attending army-ant raids).

Voice. With many species, sound plays an important role in identification. For these, we include a description of the bird's vocalization(s). Although most birds produce a variety of sounds for different purposes, we typically only describe the full song or most common call, due to space limitations. We generally describe the quality of the sound and give a transcription of the call itself in italics.

Transcribing bird sounds is a fairly subjective exercise, and we have employed several techniques to help convey how sounds are delivered. Accent marks indicate syllables that are emphasized, but not noticeably louder than the rest of the call (e.g., *huwít*). Capital letters indicate a stressed syllable that is distinctly louder (e.g., *klerEE*). The absence of punctuation marks or spaces indicates a very fast song (e.g., *bibidididididi*). Dashes indicate a slightly slower delivery (e.g., *pee-a-weee*), and commas signal an even more pronounced pause between syllables (e.g., *doy, doy, doy, doy*). And, of course, these various styles can be used together to express more complex vocalizations (e.g., *tlee-dee, teedle-doo*). When a phrase is repeated at length, an ellipsis is used to save space (e.g., *kuk-láh-kukláh-kukláh…*).

There are instances where no vocalization is included. Cotingas, for example, rarely vocalize, while other birds are so easy to see and distinguish that they can be identified without relying on sound (e.g., many waders and shorebirds).

Status and distribution. We include information about how common a species is and where it occurs geographically within the country. The elevation ranges stated in the text are not rigid; some species—especially those that feed mostly on flower nectar or fruit—will readily wander up or down slope, beyond their normal elevations, in search of food. Likewise, juvenile birds seeking a territory to claim will sometimes show up in unexpected places, though they will rarely stay long.

Also, please bear in mind that commonality can vary greatly throughout the mapped range, and is given as an approximate indication of one's likelihood of encountering the species (including by voice only) in the appropriate habitat and time of year (for migrants). The usage of terms varies somewhat from family to family; for example, a "common" raptor will never be as numerous as a "common" warbler. The terms used are the following:

- Abundant: Should be recorded daily and in appreciable numbers.
- Common: Should be recorded almost daily, though not necessarily in large numbers.
- Fairly common: Should be recorded on 50 to 75% of days.
- Uncommon: Recorded on less than 50% of days.
- Rare: Recorded on 10% or less of days.
- Casual: Not expected, but has been known to occur (typically applies to either North American migrants or wanderers from elsewhere in Costa Rica).

Range maps. Accompanying each account is a thumbnail map of Costa Rica showing the species' range within the country. These maps are rather broad-stroke representations of where each species occurs, but, taken together with the written description of a species' geographic distribution, they indicate if you are at all likely to find the species in a given area. The colors on the range map indicate the species' migratory status in the shaded area as follows:

Purple – resident. The species is known to breed in Costa Rica and is resident throughout the year. Note that, in some cases, the species may not breed in the entire mapped range (e.g., the case of altitudinal migrants).

Red – breeding migrant. The species breeds in Costa Rica, but then migrates out of the country after the breeding season.

Blue – migrant. At least part of the migratory population spends the nonbreeding season in Costa Rica (these are called winter residents), even if some portion of the population occurs as passage migrants.

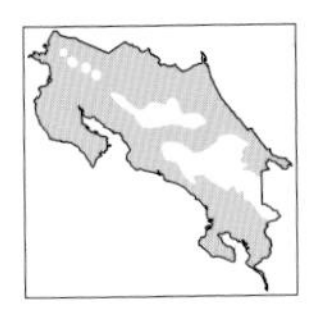

Yellow – passage migrant. The species occurs in Costa Rica en route from breeding grounds to wintering areas, and/or vice-versa. Due to the evolution of distinct migratory routes, not all passage migrant species occur both on the southbound and northbound journeys.

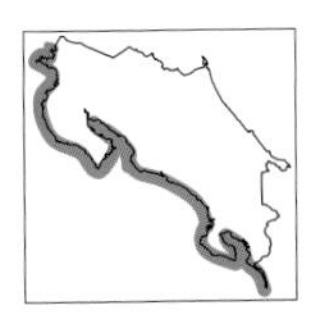

Green – casual. The species is not known to breed in the indicated area (i.e., is a migrant or wanderer from somewhere else) and is generally not expected to occur every year.

Gray – status unknown. Given lack of data, it is impossible at the time to be certain whether the species is a resident or a migrant.

As a further aid to understanding where a given species is expected to occur, the map of Costa Rica on p. 6 displays most of the geographic features and place names mentioned in the text.

Range. At the end of the account, the full extent of the species range is given (e.g., SW Arizona to W Panama).

Photographs. We include high quality photographs that exhibit the pertinent field marks of each species. A species is illustrated by multiple images when, for example, there are significant plumage differences between sexes or between adult and juvenile stages. Generally, if juvenile birds resemble adults but are simply somewhat duller, they are not shown. No attempt has been made to represent birds at the same scale. Please consult the measurements given at the beginning of each account for an idea of size.

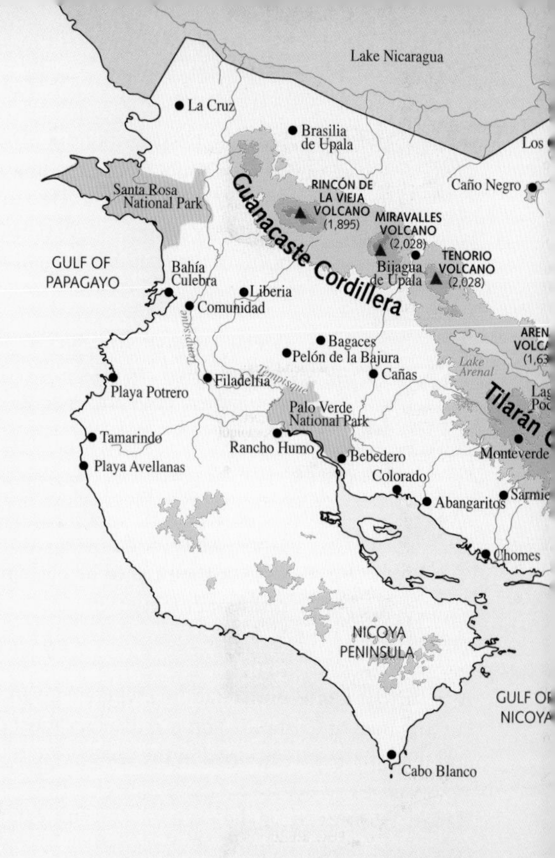
Lake Nicaragua
La Cruz
Brasilia de Upala
Caño Negro
Santa Rosa National Park
RINCÓN DE LA VIEJA VOLCANO (1,895)
MIRAVALLES VOLCANO (2,028)
TENORIO VOLCANO (2,028)
Guanacaste Cordillera
Bijagua de Upala
GULF OF PAPAGAYO
Bahía Culebra
Liberia
Comunidad
Bagaces
Pelón de la Bajura
Cañas
Lake Arenal
Filadelfia
Tempisque
Playa Potrero
Palo Verde National Park
Tamarindo
Rancho Humo
Bebedero
Monteverde
Playa Avellanas
Colorado
Abangaritos
Chomes
NICOYA PENINSULA
Cabo Blanco
Elevation (m)
2,000+ m
1,500 - 2,000 m
1,000 - 1,500 m
500 - 1,000 m
0 - 500 m
National Parks
Volcanoes
Rivers
Cities

Costa Rica

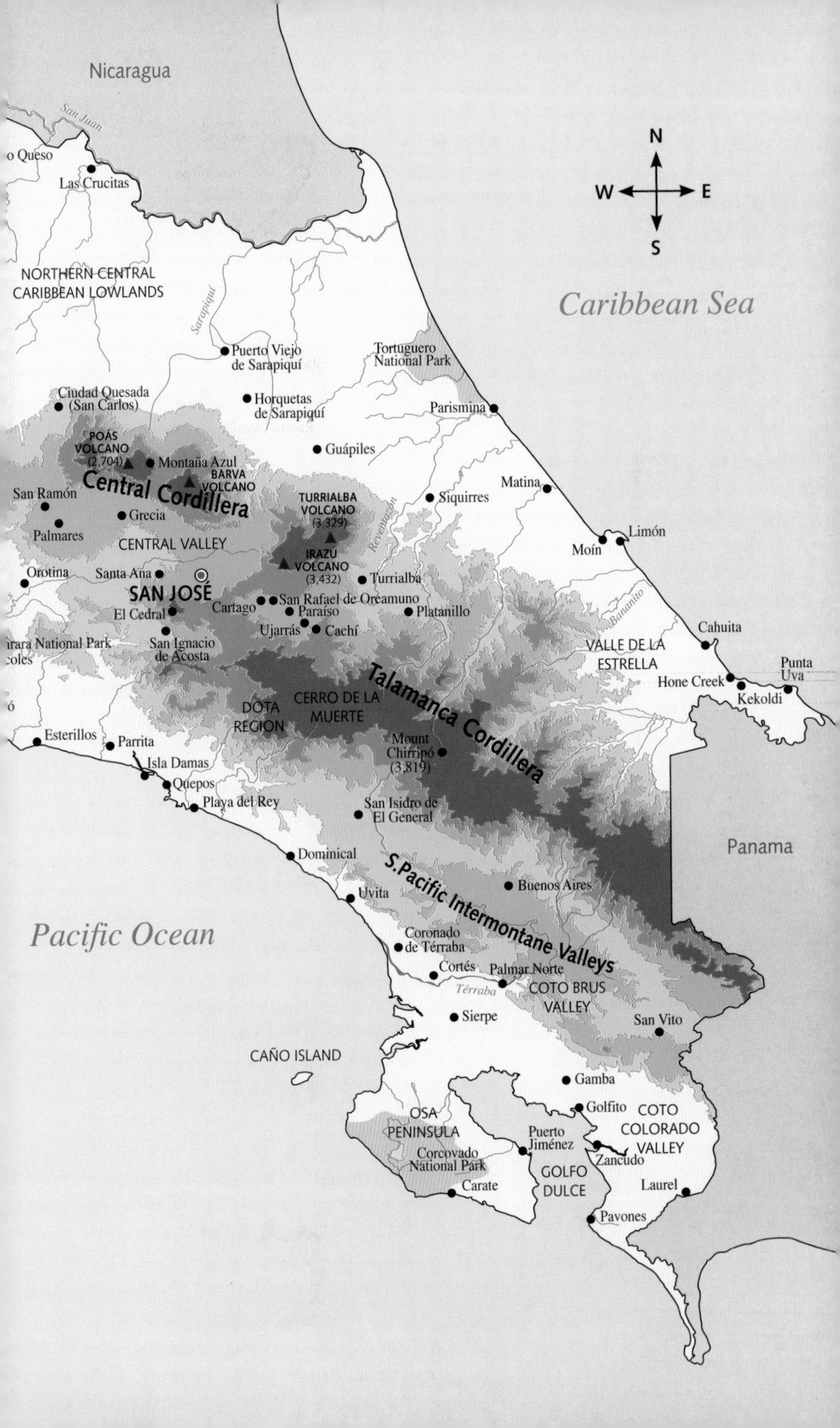

Nicaragua
San Juan
o Queso
Las Crucitas
NORTHERN CENTRAL
CARIBBEAN LOWLANDS
Sarapiquí
Puerto Viejo
de Sarapiquí
Tortuguero
National Park
Caribbean Sea
N
W
E
S
Ciudad Quesada
(San Carlos)
Horquetas
de Sarapiquí
Parismina
POÁS
VOLCANO
(2,704)
Montaña Azul
BARVA
VOLCANO
Guápiles
Central Cordillera
San Ramón
Grecia
Palmares
CENTRAL VALLEY
TURRIALBA
VOLCANO
(3,329)
Reventazón
Siquirres
Matina
Limón
Moín
IRAZÚ
VOLCANO
(3,432)
Orotina
Santa Ana
SAN JOSÉ
Turrialba
Cartago
San Rafael de Oreamuno
Paraíso
El Cedral
Platanillo
Ujarrás
Cachí
Bananito
Cahuita
San Ignacio
de Acosta
VALLE DE LA
ESTRELLA
Punta
Uva
Hone Creek
Kekoldi
Talamanca Cordillera
DOTA
REGION
CERRO DE LA
MUERTE
Esterillos
Parrita
Mount
Chirripó
(3,819)
Isla Damas
Quepos
Playa del Rey
San Isidro de
El General
Panama
Dominical
S.Pacific Intermontane Valleys
Buenos Aires
Uvita
Pacific Ocean
Coronado
de Térraba
Cortés
Palmar Norte
Térraba
COTO BRUS
VALLEY
Sierpe
San Vito
CAÑO ISLAND
Gamba
Golfito
OSA
PENINSULA
COTO
COLORADO
VALLEY
Puerto
Jiménez
Zancudo
Corcovado
National Park
GOLFO
DULCE
Carate
Laurel
Pavones

Anatomical Features

In the species descriptions, the author has used simple, nontechnical terms whenever possible. Nonetheless, in describing the various external features of birds, some technical terms become inevitable. If you are new to birding, you might want to take a few moments to familiarize yourself with the terms explained here. (See glossary, p. 14, for definitions of non-anatomical terms.)

Upperparts: Together, the crown, nape, back, rump, and uppertail coverts (the feathers that cover the base of the upper side of the tail) are known as the *upperparts*, or *dorsum*. Note: The unfeathered parts of a bird (i.e., beak, cere, orbital area, legs, and feet) are referred to as the *soft parts*.

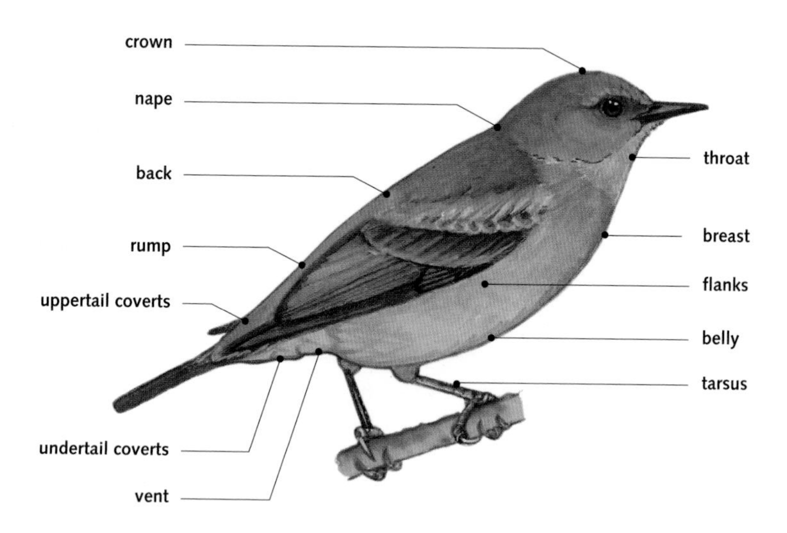

Underparts: Together, the throat, breast, flanks, belly, vent, and undertail coverts are known as the *underparts*. The vent is the feathered area around the anal opening; the undertail coverts are the feathers that cover the base of the underside of the tail. In technical parlance, the leg is referred to as the *tarsus* (plural: *tarsi*).

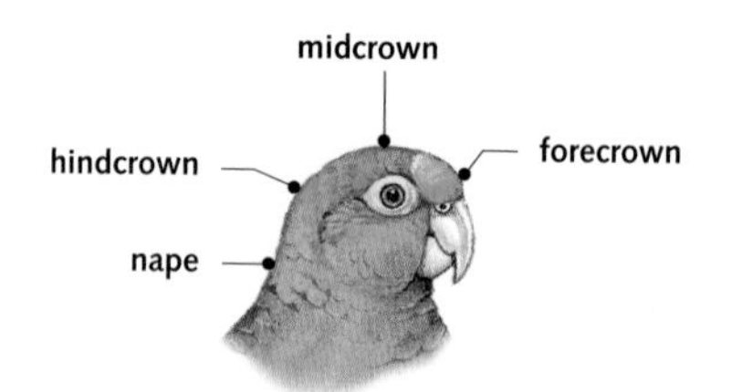

Crown: Although the entire top of the head can be called the *crown*, the anterior portion is often distinctly colored and termed the *forecrown*, or *forehead*. The terms *midcrown* and *hindcrown* refer to the center and rear of the crown, respectively. The rear portion of the head, or back of the neck, is known as the *nape*.

Wing Feathers

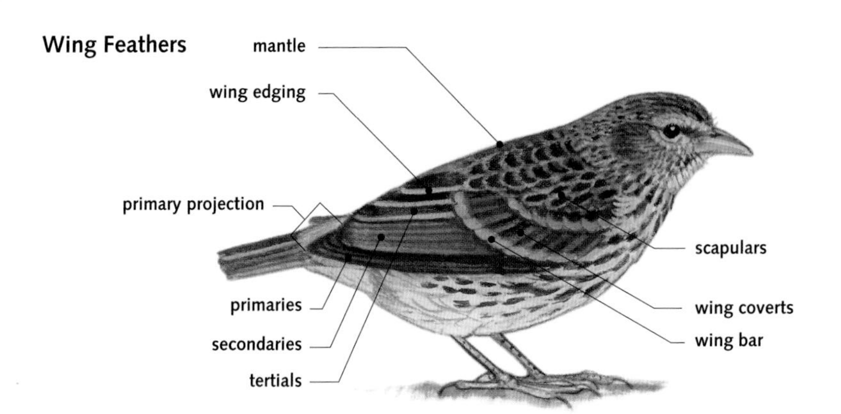

Features on the wings often help in identification. The feathers covering the juncture of the wing and body, corresponding to the shoulder, are known as the *scapulars*. Together, the scapulars and the upper back are termed the *mantle*. The several rows of feathers that cover the base of the flight feathers are the wing coverts. When the tips of these feathers are distinctly colored, they form wing bars.

The flight feathers are composed of three contiguous sets of feathers, which, from the inner to the outer portion of the extended wing, are known as *tertials* (three innermost feathers), *secondaries*, and *primaries*. When the wing is folded, these feathers stack up with the innermost of the three tertials on top and the outermost of the nine or ten primaries on the bottom.

A feature that is sometimes useful in identification is the distance that the tip of the outermost folded primary extends beyond the longest secondaries and tertials; this is known as the *primary projection*.

The term *wing edging* is used to describe a lighter color on the edge of each flight feather; this lighter color imparts a striped look to the wing.

Underwing

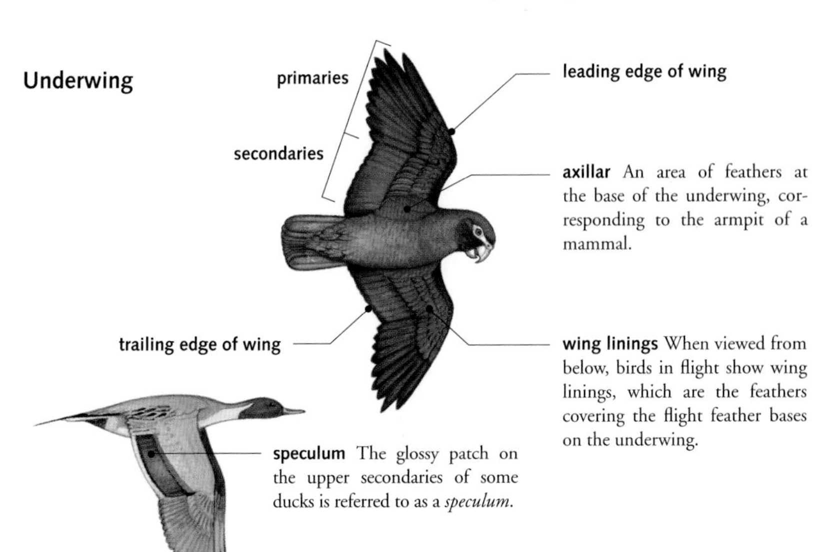

axillar An area of feathers at the base of the underwing, corresponding to the armpit of a mammal.

wing linings When viewed from below, birds in flight show wing linings, which are the feathers covering the flight feather bases on the underwing.

speculum The glossy patch on the upper secondaries of some ducks is referred to as a *speculum*.

Head: When trying to identify a bird, it is generally best to start by looking at various parts of the head. An inspection of the bill, for example, will often help you to identify the bird to the family level, if not to the species level.

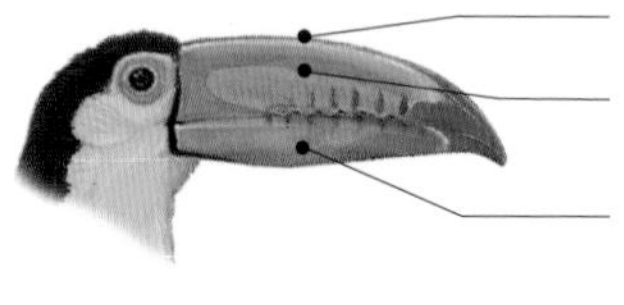

culmen The top edge of the upper mandible is called the *culmen*.

upper mandible

lower mandible The bill (or beak) comprises the upper and lower mandibles. In addition to overall bill shape and size, note any coloration differences. The term *decurved* describes a bill that has a downward curve.

cere Birds' nostrils are located on top of the base of the upper mandible. In some families (e.g., hawks, falcons, owls, pigeons, and parrots), the nostrils open through a fleshy covering called a *cere*. The color of the cere can aid in differentiating among certain species of raptors.

supraloral The area above the lore.

lore The area between the base of the bill and the eye.

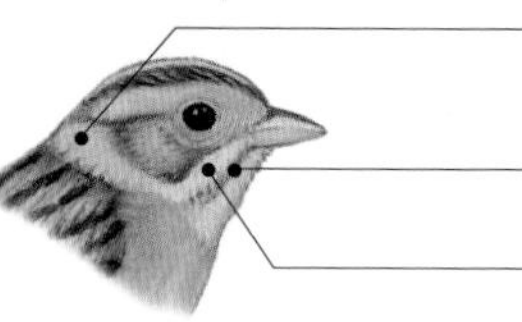

nuchal collar A distinctly colored area across the nape.

lateral throat stripe

malar stripe The area extending down from the base of the lower mandible between the cheek and throat is known as the *malar*, and when it is distinctly colored, it is called a *malar stripe*. A line between the malar and the throat is referred to as a *lateral throat stripe*. Though generally considered part of the throat, the area just below the base of the lower mandible is sometimes called the *chin*.

moustachial stripe A line of differently colored feathers just below the ear coverts.

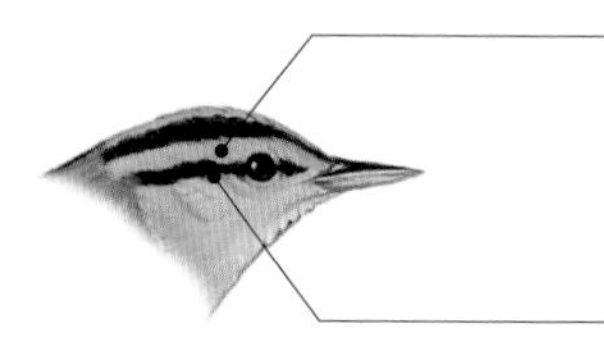

superciliary The line of feathers extending from the base of the bill to above and behind the eye (thus including the supraloral feathers) is known as the *superciliary*. In many species, these feathers are of a different color than the contiguous head feathers, making them very useful in species identification.

eye line When the lore and the postocular stripe (see next page) are of the same color and set off from the surrounding feathers, they form an eye line.

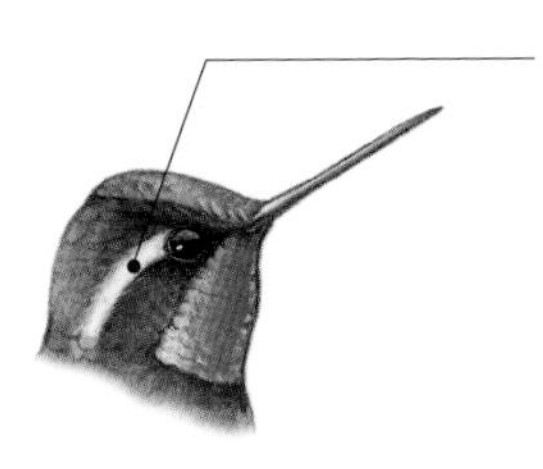

postocular stripe

When a differently colored line of feathers extends behind the eye, it is referred to as a *postocular stripe*. Note that some birds show only a postocular spot.

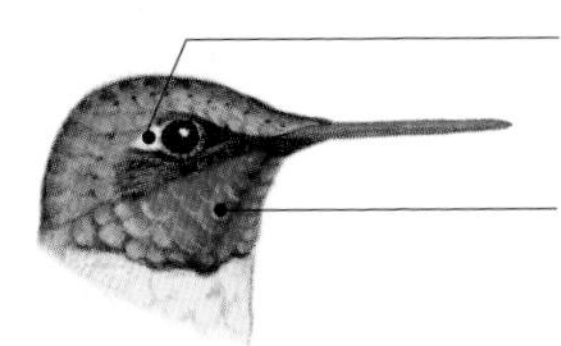

postocular spot

gorget The highly iridescent throat feathers on certain species of hummingbirds are known as the *gorget*.

eye ring

Most species of birds have a narrow ring of very short feathers around the eye. When these feathers are of a different color than the surrounding ones, they form an eye ring. In some species, the area around the eye (and sometimes beyond) is featherless. This bare area is referred to as *orbital skin*, or, when it forms a circle, as an *orbital ring*.

orbital skin

spectacles (formed with the lore)

Sometimes an eye ring is the same color as either the lore or a postocular stripe; the resulting pattern is called *spectacles*.

spectacles (formed with a postocular stripe)

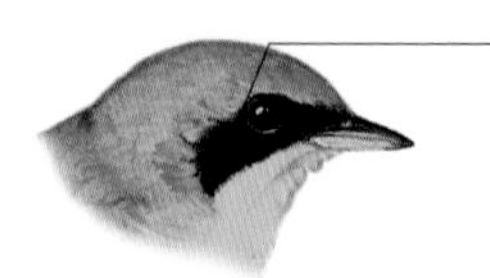

mask The term *mask* is often used when the area around the eye is a darker color—usually black—than the rest of the head. More often than not, the mask includes the lore and ear coverts, but the term may be used in reference to little more than the lore.

ear coverts & ear patch The ear coverts, or cheek, are a distinct group of short feathers that cover the ear opening, which is located just below and behind the eye. When the feathers of the ear coverts, or sometimes just those along their outer border, are of a different color and contrast with the feathers surrounding them, the result is an ear patch.

crown patch Some species, including many of the New World flycatchers, have a colorful crown patch of erectile feathers that is generally kept concealed beneath the crown feathers.

crest When most or all of the crown feathers are capable of being raised, they are termed a *crest*.

hood The term *hood* has variable usage, though it generally indicates the crown, nape, and lower throat (excluding the forehead, eyes, and chin).

half-hood A half-hood is made up of the crown and cheeks, but does not include the throat.

Tail Feathers

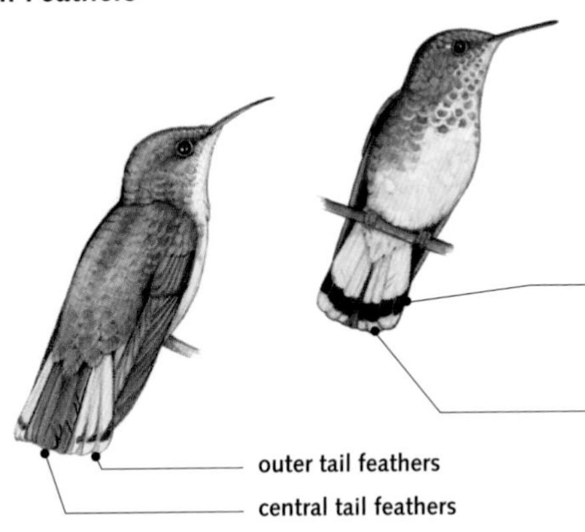

subterminal band A distinctly colored band above the tail tip is called a *subterminal band.*

terminal band The tip of the tail is composed of the ends of all the tail feathers, and if they are of a distinct color they form a terminal band.

Tail Shapes: In addition to coloration or markings, note the shape of the tail, especially paying attention to whether the tip is rounded, notched, forked, square, tapered, or pointed. Also note the proportion of the tail length with respect to the body.

Glossary

See Anatomical Features (p. 8) for additional definitions.

allopatric. Ocurring in different geographic areas.
barred. Marked with lines that are horizontal in relation to the upright axis of the bird.
basal. Pertaining to the part of the beak, wing, or tail nearest to the base. See **distal**.
breeding plumage. The distinctive, and usually brightly colored and/or boldly marked, plumage that a bird molts into just prior to the breeding season. Many North American migrant species appear in Costa Rica in these plumages in March and April. See **nonbreeding plumage**.
brood parasite. A species that deposits its egg(s) in the nest(s) of other species, thus freeing itself from all nesting chores.
buffy. Having a creamy, light-brown color.
canopy. The uppermost level of a forest.
commensal. Describes birds that live in close proximity to human society and benefit from it either by feeding on human-generated refuse or by nesting on human-made structures.
congeneric. Belonging to the same genus.
congeners. Species belonging to the same genus.
conspecific. Belonging to the same species.
cosmopolitan. Found on all six continents (Antarctica excluded).
covey. A family group of quails.
CR. Acronym denoting *Costa Rica*.
cryptic. Describing a plumage pattern and/or behavior that makes a bird blend in with its surroundings.
distal. Pertaining to the part of the beak, wing, or tail farthest from the base. See **basal**.
dry forest. Forests of the northern Pacific region and western Central Valley that are dry from November to April.
dusky. Dark-colored, usually dark grayish; but as with most colors employed in bird names, usage is somewhat varied.
endemic. Occurring only within a limited geographic area.
epiphyte. Any species of plant evolved to grow on another plant (but that is not parasitic), such as many of the mosses, ferns, orchids, and bromeliads.
eye-shine. The color reflected from a bird's eye when a beam of light is directed at it. The observer must be directly in line with the beam to see this.

fasciated. Marked with fine, wavy barring.
first-winter bird. Refers to gulls during the first winter after hatching (approximately September to April). Gulls take two or more years to reach adult nonbreeding plumage.
flotsam. Accumulated debris at the high-water line or floating in the open ocean.
foothills. Indicates the elevation range of approximately 1,300 to 2,600 ft (400 to 800 m); this range is somewhat higher in the southern half of Costa Rica.
forest edge. The transitional zone between a forested area and a more open area, which is often very productive for birding. Forest canopy species sometimes come down to much lower levels along this "vertical canopy."
gallery forest. Forest that grows along a river or stream.
garden. A catchall term referring to any human-altered environment that still contains some bushes and trees.
highlands. Indicates the upper portions of mountain slopes, from approximately 4,900 to 12,500 ft (1,500 to 3,800 m).
humid forest. Forest that remains largely evergreen, but receives less rain than **wet forest**; one indicator of this type of forest is the paucity of epiphytic plants.
irruptive. Denotes a population that experiences periodic peaks of abundance.
jizz. In birding parlance, the general visual impression of a family, genus, or species based on overall size, appearance, and behavior; a subjective but very real tool in the identification process.
lowlands. The elevation range from sea level to approximately 1,600 ft (500 m).
mature forest. Forest that has not been disturbed by human activities or cataclysmic natural forces in recent times.
middle elevations. Indicates the elevation range from approximately 2,600 to 5,900 ft (800 to 1,800 m); this range is somewhat higher in the southern half of Costa Rica.
montane. Pertaining to cooler areas of middle elevations and highlands.
morph. One of two or more different color forms in which a species occurs. Since an individual bird remains in a given color morph for its entire life, the author prefers this term over the often used *phase*, which could imply that an individual's coloration is not permanent.
neotropical. Pertaining to the tropics of the Americas.
nonbreeding plumage. The drabber, and often less obviously marked, plumage that a bird shows when it is not breeding. Many North American migrant species appear in Costa Rica in these dull plumages. See **breeding plumage**.

NP. Acronym denoting *National Park.*
ochraceous. A warm blend of yellow, orange, and brown tones.
pantropical. Pertaining to the tropics around the globe.
paramo. A habitat found above timberline that is composed of low vegetation.
passerine. Any of the so-called perching birds or songbirds. More technically, any species in the order Passeriformes.
pelagic. Pertaining to the open ocean.
precocial. Refers to birds that are relatively mobile upon hatching.
race. Synonymous with **subspecies**; usually mentioned when there is some notable difference between two such groups.
raptor. A bird of prey (i.e., a hawk, eagle, falcon, or owl).
rufous. A reddish-brown color.
sally. A flight made in pursuit of a prey item.
salt ponds. Shallow depressions (usually constructed in areas that were originally mangrove swamps) that are filled with sea water, which is then left to evaporate, leaving salt crystals. Many shorebirds utilize these areas at high tide, both to forage in the ponds and rest on the dikes that separate them.
second growth. Vegetation composed of fast growing trees, vines, and shrubs that colonize areas affected either by natural causes (fire, flooding, volcanic eruptions, landslides, etc.) or cutting by humans.
shorebird. Refers principally to members of the sandpiper, plover, and stilt families, most of which are typically encountered near coastal areas.
streaked. Synonymous with **striped**.
striped. Marked with lines that are vertical in relation to the upright axis of the bird.
subspecies. Synonymous with **race**.
sympatric. Occurring in the same geographic area.
timberline. The upper elevation limit of tree growth; in Costa Rica this occurs at about 10,200 ft (3,100 m).
understory. The lower level of forest habitat, from ground level to about 13 ft (4 m) up.
vermiculation. A pattern of wavy lines.
wader. Refers principally to members of the heron, ibis, and stork families, which are typically encountered in wetlands, where their long legs enable them to wade in shallow water.
wet forest. Forest that receives enough annual moisture (at least 8 ft [2.5 m]) for most of the plant species to remain evergreen; wetter than **humid forest**.

Species Accounts and Photographs

A note on the photographs: Please keep in mind that birds are not presented at scale—even birds on the same page. In almost all cases, if a species is represented by just a single photograph, then that species shows no (or little) variation in plumage; in cases where the adult and juvenile differ in appearance—and where only a single photograph is included—the photograph is of an adult.

Abbreviations: The letters *N, E, S, W* are used to indicate cardinal directions, as are combinations of the same (SW = southwest, for example).

CA = Central America
CR = Costa Rica
NA = North America
NP = national park
SA = South America

Order of presentation of species: For reasons of design, in a handful of cases the order in which families (and/or species) appear differs from that of the second edition of *The Birds of Costa Rica:* A Field Guide (Garrigues and Dean).

Tinamous (TINAMIDAE). A neotropic family whose closest relatives are the rheas, kiwis, and the ostrich. Unlike their distant kin, tinamous are capable of flight, though only for short distances. Their diet is opportunistically varied with fallen fruits and seeds, as well as captured arthropods and small frogs and lizards. Forest-dwelling species typically nest between the buttresses of a tree; eggs are laid on the ground. The male is responsible for incubating the eggs and caring for the precocial chicks during their first few weeks. World: 47, CR: 5

Great Tinamou
Tinamus major

17 in (43 cm). The country's largest tinamou. This species inhabits mature wet forest, where its far-carrying tremulous whistle is a characteristic sound. The combination of gray legs and black barring on the hindparts distinguishes it from other tinamous. The earthen tones of its plumage provide effective camouflage when it stands motionless on the forest floor. Sought after as a game bird, it can be quite wary, but in protected areas is often fairly tame. If threatened, it prefers to walk away, but if startled, will burst into flight with an explosive whirring of wings. Fairly common in lowlands; decreasingly common up to 5,600 ft (1,700 m). Range: SE Mexico to Amazonian Brazil.

Little Tinamou
Crypturellus soui

9 in (23 cm). The widest ranging and smallest tinamou in CR, though difficult to see given its predilection for dense second growth. If glimpsed, note its unpatterned plumage and dull-yellow legs. It gives a short, clear whistle that rises, then falls slightly, recalling a horse's whinny. Fairly common in wet and humid regions; to 4,900 ft (1,500 m). Range: S Mexico to E Brazil.

Curassows, Chachalacas, Guans (CRACIDAE). The chachalacas, guans, and curassows constitute a neotropical family of birds related to chickens and turkeys, although they are largely arboreal—even the heavy curassows roost and nest in trees. Unlike the members of most other families related to chickens, the cracids typically lay only two to four eggs per nest. The precocial young are fed regurgitated plant material by their parents. World: 54, CR: 5

Gray-headed Chachalaca

Ortalis cinereiceps

20 in (51 cm). The most widespread chachalaca in CR. Its rufous primaries distinguish it from the otherwise very similar Plain Chachalaca *O. vetula* (not illustrated) of the NW dry forests. Gregarious, it travels through second growth and along forest edges in groups of up to a dozen or more, feeding on fruits and leaves. Unlike other members of the genus, it does not give a raucous call, but produces a variety of sounds varying from soft, high notes to rather loud clucking. Fairly common in wet lowlands and foothills; to 3,900 ft (1,200 m). Range: E Honduras to NW Colombia.

Crested Guan

Penelope purpurascens

35 in (89 cm). Nearly as large as a Great Curassow (p. 20), but weighing less than half as much, this is one of the largest arboreal species. Pairs or small groups feed on fruits and tender foliage in middle and upper levels of mature forest, advanced second growth, and forest edges; rarely comes to the ground except to feed on fallen fruit. The loud wing-rustling flight displays (and far-carrying *klah-klah-klah*…) are heard at dawn and dusk. Still fairly common in protected areas throughout the country; to 5,900 ft (1,800 m) in southern Pacific, elsewhere to 3,900 ft (1,200 m). Range: Mexico to SE Ecuador and E Venezuela.

Great Curassow
Crax rubra

male

36 in (91 cm). The unique curly crest feathers set apart both sexes of this dimorphic species. Some females in northern half of CR are conspicuously barred. Generally encountered foraging on the ground inside mature forest, but will readily fly up into trees, both to elude predators as well as to roost. Males emit an almost inaudibly deep *uuuhm*. Uncommon to rare in protected areas throughout the country; to 3,900 ft (1,200 m). Range: E Mexico to W Ecuador.

female

female barred morph

Black Guan
Chamaepetes unicolor

24 in (61 cm). The only member of its family in most of its range, it is easily identified by its blue facial skin and reddish legs. Individuals or pairs can be found in fruiting trees in mature highland forest. Most often heard sound is a startling wing-rattling, recalling a muffled burst of machine-gun fire. Fairly common in protected areas of highlands, uncommon to rare elsewhere; from 3,600 ft (1,100 m) to timberline. Range: CR and W Panama.

New World Quail (ODONTOPHORIDAE). Formerly considered members of the same family as pheasants and grouse, the New World quails are short, chunky birds with strong feet for scratching in leaf litter for tubers, grubs, and fallen fruit and seeds. Their dark-brown plumage provides excellent camouflage against the shadowy forest floor, which, combined with a wary nature, results in their being heard far more often than seen. Nests are built on the ground and typically contain four or five eggs (up to ten or more in bobwhite nests). World: 33, CR: 7

male northern Pacific race

Crested Bobwhite
Colinus cristatus

8 in (20 cm). The only quail in most of its CR range, it inhabits grasslands and brushy fields in drier regions, where coveys of up to a dozen or more birds forage in dense cover. Whistles a characteristic *whit, whit, wEEit*. Northern Pacific race is common in northern Pacific and uncommon in Central Valley; to 4,900 ft (1,500 m). Southern Pacific race is rare in lowlands on mainland side of Golfo Dulce; also reported from hills south of San Isidro de El General. Range: S Guatemala to Colombia and N Brazil.

Spotted Wood-Quail
Odontophorus guttatus

10 in (25 cm). No other CR quail combines a streaked black-and-white throat and spotted underparts. Small coveys feed on the ground in mature highland forest, thickets, second growth, and forest openings. At dawn, birds repeat a loud *hu-wit, hu-wit, hu-wit-tsú*. Uncommon on Central and Talamanca Cordilleras; from 3,300 ft (1,000 m) to timberline on Pacific slope, and from 4,900 ft (1,500 m) to timberline on Caribbean slope. Range: SE Mexico to W Panama.

Sungrebe (HELIORNITHIDAE). This pantropical family has three species, with one each in Africa, SE Asia, and the New World. Lobed toes and colorful legs and feet are shared features, together with a largely aquatic lifestyle—although these birds are also perfectly suited for walking on land, perching in vegetation, and flying. Unlike grebes and coots, which also have lobed toes for powered swimming, the "finfoots," as they are often called, do not dive. All three species are fairly omnivorous, though insects feature prominently in the diet. Pairs construct shallow stick nests amid vegetation in branches over water. World: 3, CR: 1

Sungrebe
Heliornis fulica

breeding female

11 in (28 cm). The black-and-white pattern of the head and neck sets it apart from local ducks and grebes. In breeding condition, the female acquires tawny cheeks and a coral-pink color to the eyelids and upper mandible. As a unique adaptation among birds, the male can carry a chick—even in flight!—in a fold of skin that forms a pocket under the wing. Favors slow-moving rivers and streams with banks covered by overhanging vegetation, from which it gleans arthropods, frogs, and small lizards. Fairly common in Caribbean lowlands; rare in Golfo Dulce region. Range: E Mexico to NE Argentina.

male/nonbreeding female

Grebes (PODICIPEDIDAE). Evolved for an aquatic lifestyle, grebes have lobed toes on legs set far to the rear of their bodies. While providing excellent propulsion in water, this arrangement renders them quite awkward on land. Hence they seldom leave the water except to tend the nest, which consists of layers of plant material anchored to floating vegetation. They feed on fish, crustaceans, and insects that are typically caught and consumed underwater. World: 22, CR: 3

Least Grebe

Tachybaptus dominicus

breeding

nonbreeding

9 in (23 cm). The small size, dark plumage, and yellow eye distinguish this species from other duck-like birds. Individuals or pairs can be found on freshwater ponds and reservoirs, and even in temporary pools formed by rainwater, which are often utilized for nesting. Feeds primarily on aquatic invertebrates. Common in the Central Valley and northern Pacific; fairly common elsewhere, but rare in Tortuguero NP region; to 4,900 ft (1,500 m). Range: S Texas to N Argentina.

Ducks (ANATIDAE). Ducks, geese, and swans are among the most familiar birds due to their worldwide distribution and the domestication of several species since ancient times. They are well adapted to aquatic habitats: Wide bodies displace their weight in water, strong legs and webbed feet help propel them, and a gland at the base of the tail produces waterproofing oil with which they preen their feathers. Despite the distinctive bill shape—which might suggest specialization on one type of food—diet is rather varied throughout the family and includes plant matter, insect larvae, fish, and molluscs. World: 164, CR: 21

Black-bellied Whistling-Duck

Dendrocygna autumnalis

21 in (53 cm). The most common and widespread resident duck in CR. The adult is easily distinguished by its coral-pink bill and pink legs. These soft parts are dark gray on the juvenile; however, both juvenile and adult show a distinctive white wing stripe in flight. Although often alert and active by day, this species forages mostly at night, feeding on seeds, shoots, and invertebrates. Nests can be placed either on the ground or in cavities in trees. The common name is derived from its typical call: a whistled *whit-whit-wee-wee-wee*. Abundant in Palo Verde NP and Caño Negro regions, fairly common elsewhere; to 4,900 ft (1,500 m). Range: S Texas to N Argentina.

adult

In flight, all plumages show a white wing stripe.

male

Muscovy Duck
Cairina moschata

Male 33 in (84 cm); female 25 in (64 cm). By far, the largest duck in CR, with males weighing in at 6.6 lbs (3 kg). Males have red fleshy protruberances around the face. In flight, adults show large white patch on forewing; juvenile has a small patch. Forages for seeds in open areas, and opportunistically takes insects and frogs. Prefers wooded areas—forest streams, swamp forest, or mangroves—for nesting since eggs are laid in a tree cavity. Interestingly, the precocial, but flightless, fledglings generally survive the fall to the ground or water after they hatch and leave the nest. Fairly common in Palo Verde NP region, rare elsewhere in lowlands. Range: S Mexico to NE Argentina.

Blue-winged Teal
Anas discors

breeding male

female/nonbreeding male

15 in (38 cm). This migrant species is the most numerous duck in CR, with hundreds, or even thousands, of individuals present at suitable wetland sites. The male in breeding plumage is readily identified by the white crescent facial mark. Females and nonbreeding males are brownish with grayer faces and a thin, dark eye line. The light-blue wing coverts are visible in flight. With its bill, dabbles on the water surface for plant matter and aquatic insects. Abundant migrant in Palo Verde NP and Caño Negro regions and in Central Valley, uncommon elsewhere; to 4,900 ft (1,500 m); from Sept to April (a few present year-round). Range: Breeds in N NA, winters south to N Argentina.

Cormorants (PHALACROCORACIDAE). The members of this cosmopolitan family live in coastal and freshwater habitats, where they feed on fish, as well as amphibians and aquatic invertebrates. A cormorant catches prey in its hook-tipped bill while swimming underwater, using its strong, webbed feet to propel itself. When swimming on the surface, birds often remain partially submerged, showing only their head, neck, and back. They also spend a lot of time time perched on rocks or logs, or in trees, with their wings outstretched to dry their permeable plumage. World: 40, CR: 1

Neotropic Cormorant

Phalacrocorax brasilianus

juvenile

adult

26 in (66 cm). The only cormorant in CR, it can be told apart from the Anhinga by its hook-tipped bill and entirely dark wings. Roosts and nests colonially in trees overhanging water, where it makes noisy, pig-like grunts that give rise to the CR common name *pato chancho* (pig duck). Common from lowlands to middle elevations; to 4,900 ft (1,500 m). Favors slow-moving bodies of water. Range: S US to Tierra del Fuego.

Anhingas (ANHINGIDAE). With just one species in the New World and three in the Old World, the anhinga and darters inhabit freshwater and coastal habitats in tropical and warmer temperate regions. Closely related to the cormorants, they also catch fish and other prey underwater; however, they do so by spearing the prey on their pointed bill. Their plumage absorbs even more water than that of cormorants—allowing them to submerge effectively—and thus they also are frequently seen on perches just out of the water with wings stretched as they dry themselves. Unlike their relatives, they do not form large congregations for roosting and nesting. World: 4, CR: 1

male

Anhinga

Anhinga anhinga

35 in (89 cm). The long, pointed bill and whitish markings on the wings distinguish it from the Neotropic Cormorant. Female and juvenile have buffy head and neck. Typically swims with just the neck and head above water. Also soars regularly. The nest is a platform of sticks, lined with fresh leaves, usually in a branch of a tree overhanging water. Although fairly common, it is typically less numerous than the sympatric Neotropic Cormorant. Range: SE US to N Argentina.

female

An anhinga drying its wings.

Pelicans (PELECANIDAE). The large size, long bill, and unique extendible gular pouch make pelicans easy to recognize. Awkward on land, they are proficient swimmers and graceful fliers. Fish constitute the bulk of their diet and, especially when raising young, they consume large quantities. This, along with their gregarious habits, limits their presence to areas with sufficiently abundant food resources. They are colonial breeders; some species build nests in trees, while others are ground nesters—the latter generally choose sites such as islands that offer protection from terrestrial predators. World: 8, CR: 2

Brown Pelican

Pelecanus occidentalis

juvenile

47 in (116 cm). Nonbreeding adult has a white head and neck, which could cause confusion with the very rare migrant American White Pelican *P. erythrorhynchos* (not illustrated), but Brown Pelicans have grayish (not white) back feathers. When feeding, it plunges headlong into the water, opening its bill just before impact to engulf the small fish it has in its sights. Numerous individuals soar in a line or V-formation. Often rests on sandbars or perched in trees, especially mangroves. Though present on the Caribbean coast, it is far more common on the Pacific coast, where hundreds can often be seen in a single day. Range: Mid-Atlantic US to N SA; California to N Peru.

breeding adult

Frigatebirds (FREGATIDAE). Members of this very distinctive pantropical marine family are characterized by having the greatest ratio of wing area to body weight of all the world's birds. Thus, they soar freely and at length, as well as engage in spectacular displays of maneuverability and aerial prowess. They feed largely on fish snatched from the surface, but also pursue other birds, pestering them until they forfeit their prey. Offal from fishing boats is another important food source. A perched male frigatebird displays by inflating his bright red gular pouch, but this is rarely observed on birds away from nesting sites. They roost and nest colonially in trees on small offshore islets. World: 5, CR: 2

Magnificent Frigatebird

Fregata magnificens

male

juvenile

40 in (102 cm). The only frigatebird likely to be seen from the mainland. Female has dark head and white breast; juvenile has white head and breast. Common along Pacific coast, fairly common along Caribbean coast; occasionally, an individual can be seen soaring far inland. Range: Tropical W Atlantic and E Pacific oceans.

Boobies (SULIDAE). The gannets and boobies are oceanic species with a characteristic "four-pointed cross" silhouette in flight—having pointed beak, wings, and tail. They feed primarily on fish and squid, which they catch underwater or at the surface on powerful dives. They roost and breed colonially. In some places, the concentrations of their accumulated guano, together with that of cormorants and pelicans, have resulted in a lucrative fertilizer trade for local human populations. The name *booby*, from the Spanish *bobo* (stupid), is in reference to their lack of fear of humans. World: 10, CR: 5

Brown Booby
Sula leucogaster

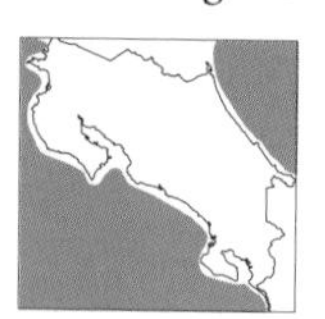

27 in (69 cm). The smallest and perhaps most widespread and numerous of all the boobies, this is the most commonly seen booby on both coasts of CR, though it is more numerous on Pacific. Males of the Pacific race have pale-gray heads. Juvenile is gray-brown with a lighter gray-brown belly. Roosts and breeds on rocky islets. Nests on the ground and throughout the year. Rarely comes very close to shore. Range: Circumtropical.

male Caribbean race/female (both races)

juvenile

male Pacific race

Gulls, Terns, Skimmers (LARIDAE). The members of this cosmopolitan family are generally associated with coastal environments, though there are species that occur far inland and others well out to sea. Most species have complex molt patterns and take three or more years to reach adult plumage. Identification is often aided by soft-part colors. Gulls are heavier billed, on average, than terns and typically are opportunistic feeders that swim well, but seldom dive. Terns generally dive for fish or catch insects in flight; they rarely swim. Skimmers are different enough to sometimes be considered in a family of their own. Most lay their eggs in a scrape on the ground, though some species construct a nest of grass or other material, and a few nest in trees. World: 99, CR: 31

Laughing Gull

Leucophaeus atricilla

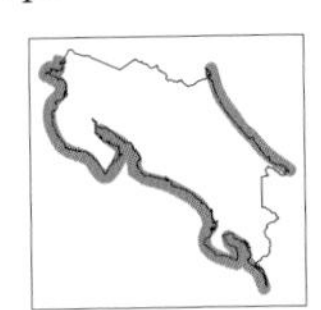

breeding adult

16 in (41 cm). Of the ten gull species reported in CR, this is the one most likely to be encountered. Opportunistically scavenges along tideline and around ports and fishing boats; rests on sandbars and on dikes that crisscross salt ponds and commercial shrimp ponds. Common migrant along both coasts (rarely inland) from Sept to May; juveniles present year-round. Range: Breeds coastally from Maine and SE California south to West Indies and CA, winters south to Peru and N Brazil.

nonbreeding adult

first winter

first summer

First winter bird shows complete black tail band.

Royal Tern
Thalasseus maximus

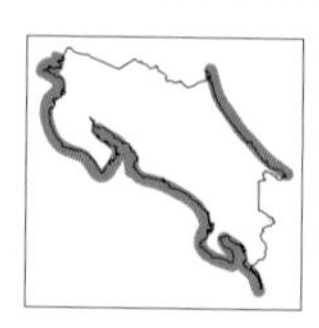

breeding

20 in (51 cm). By far, the most common tern in CR. Note the stout orange-yellow bill. Feeds offshore, plunge-diving for fish; rests on sandbars, also on dikes that criss-cross salt ponds and commercial shrimp ponds. Common migrant along both coasts from Sept to June; juveniles present year-round. Range: Breeds coastally from E US and SE California south to Uruguay, winters south to Argentina; also breeds coastally from Mauritania to Guinea, winters south to Namibia.

nonbreeding (has white forehead and midcrown)

Sandwich Tern
Thalasseus sandvicensis

breeding

16 in (41 cm). Readily distinguished by the pale tip on an otherwise black bill. Generally the most numerous tern in CR after Royal Tern, with which it associates. Fairly common migrant on both coasts, from Sept to May; juveniles, present June to Aug, are more common on Pacific coast. Range: E US to N Argentina, northern birds winter south to central SA; breeds from Europe to Caspian Sea, winters south to Africa, India, and Sri Lanka.

nonbreeding

Black Skimmer
Rynchops niger

18 in (46 cm). The name derives from the foraging technique of flying low over calm water with the tip of the noticeably longer lower mandible skimming the surface. When a prey item is felt, the mandibles immediately close, seizing the meal. Usually seen resting on sandbars with gulls and terns. Fairly common NA migrant in Gulf of Nicoya, rare elsewhere along the Pacific coast; from Sept to May. Rare summer resident on the Pacific coast, from June to Aug. From May to Oct occasional birds from SA are present and can be told apart by their entirely dark tail feathers. Range: Breeds from Massachusetts and S California to S Mexico, winters to Panama; resident in SA to N Argentina.

Sunbittern (EURYPYGIDAE). Outwardly similar to the herons and bitterns (hence the confusing English common name), the Sunbittern is actually different enough from all other birds to merit being placed in a family by itself. The overall gray coloration provides camouflage amid river rocks, but when the wings are spread they reveal a spectacular sunburst pattern that is used primarily as a threat display. Aquatic invertebrates and small vertebrates form the bulk of the diet. Pairs build a nest of sticks and leaves held together with mud and usually placed on a horizontal branch over water. World: 1, CR: 1

Sunbittern
Eurypyga helias

18 in (46 cm). The long, sharp bill and fairly long neck and legs, together with its fondness for water, all combine to suggest a type of heron; however, the blackish head with a white superciliary and long white malar stripe is diagnostic. Forages individually in two distinct habitats: rushing, mountain streams and lowland swamp forests. Emits a far-carrying, forlorn, rising whistled note lasting about one second. Uncommon in wet lowlands and foothills of Caribbean and southern Pacific (north to about Quepos); rare along streams on Pacific slope of the Guanacaste, Tilarán, and Central Cordilleras; to 4,900 ft (1,500 m). Range: S Mexico to E Bolivia.

Herons, Egrets, Bitterns (ARDEIDAE). Herons, with few exceptions, are characterized by having long legs, neck, and bill. Their long legs allow them to wade in shallow water in search of fish. The arrangement of their neck vertebrae enables them to strike at prey with blinding speed. And the long, sharp bill impales the unfortunate object of attack. They also take small vertebrates and a wide variety of arthropods. Most species nest colonially in trees. Species that breed at higher latitudes migrate to the tropics. World: 64, CR: 19

Bare-throated Tiger-Heron

Tigrisoma mexicanum

adult

30 in (76 cm). The common and widespread tiger-heron of lowlands on both sides of the country, rarely above 1,600 ft (500 m). If the bare yellow throat skin cannot be seen, note the black crown and gray cheeks of the adult. Without seeing the throat, juveniles are difficult to tell apart from other juvenile tiger-herons. Forages at the edge of slow-moving water and can be seen on rivers and streams, as well as in mangroves, marshes, and even roadside ditches. Nests solitarily in the crown of a large tree. Utters a deep, rolling *wowrrh*. Range: N Mexico to NW Colombia.

juvenile

Fasciated Tiger-Heron
Tigrisoma fasciatum

25 in (64 cm). The only tiger-heron likely to be found on rocky, fast-flowing streams. The fine barring on the cheeks and crown of the adult readily distinguishes it from the adult Bare-throated Tiger-Heron. Juveniles of both species are very similar. Usually seen singly, often standing motionless as it watches for prey from a rock in midstream or on the bank. Uncommon in Caribbean foothills, as far north as Miravalles Volcano; from 300 to 3,000 ft (100 to 900 m). Rare in Pacific foothills between Quepos and Uvita, and in interior of the Osa Peninsula. Range: CR to N Argentina.

Great Blue Heron
Ardea herodias

46 in (117 cm). The largest heron in CR. Readily told apart by head pattern and grayish coloration. Usually seen singly, though often with other heron and egret species. Found in almost any open wetland habitat. A fairly common and widespread migrant, mostly from Sept to May, but some individuals present year-round; to 4,900 ft (1,500 m). Range: Breeds in N NA, winters south to N SA.

Great Egret
Ardea alba

39 in (99 cm). Larger than any of the other white egrets and herons, of which only the much smaller Cattle Egret (p. 39) also has a yellow-orange bill. Particularly majestic in flight, with long neck typically held recoiled in an S-shape. Found in almost any open wetland habitat. Common and widespread; to 4,900 ft (1,500 m). Resident population joined by migrants from Sept to April. Range: Cosmopolitan.

Snowy Egret
Egretta thula

24 in (61 cm). The slender, black bill and black legs with yellow feet impart a handsome aspect to this medium-sized white egret. Seen singly as well as in groups. Found in almost any open wetland habitat, and along wooded streams. Roosts and nests communally. Common and widespread; to 4,900 ft (1,500 m). Resident population joined by migrants from Sept to April. Range: US to central Argentina.

Little Blue Heron
Egretta caerulea

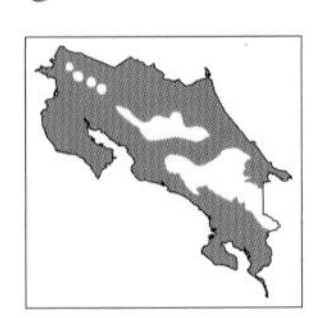

24 in (61 cm). Both the slate-blue adults and the white juveniles can be told apart from any similar species by two-tone bills that have a blue-gray base and a dark tip. Young birds molting into adult plumage exhibit piebald patterns. Found in almost any open wetland habitat, and along wooded streams. Common and widespread migrant, from Sept to April; uncommon summer resident, from April to Sept; to 4,900 ft (1,500 m). Range: US to S Brazil.

adult

juvenile

Tricolored Heron
Egretta tricolor

adult

26 in (66 cm). Recalls adult Little Blue Heron (p. 37), but note white stripe down neck and white belly. Forages singly and often quite actively, opening wings and chasing prey. Found in almost any open wetland habitat, though generally prefers areas near coasts. Fairly common and widespread migrant, mostly from Sept to May, but some individuals present year-round; to 4,900 ft (1,500 m). Range: S US to S Peru.

juvenile

Cattle Egret
Bubulcus ibis

Nonbreeding adult perched on a horse.

breeding adult

20 in (51 cm). The smallest of the white egrets, the yellow-orange bill sets it apart from all but the much larger Great Egret (p. 36). Breeding adult acquires pale-rusty feathers on crown, breast, and back. Found in open wetland habitats, as well as in pastures, where it follows livestock, grabbing the insects that the animals stir up; also follows tractors. Roosts and nests colonially. Since arriving in CR in 1954, has become common and widespread; to at least 7,200 ft (2,200 m). Range: Cosmopolitan.

Green Heron

Butorides virescens

On breeding adult male, note deep-orange legs.

On female and nonbreeding male, note yellow-orange legs.

18 in (48 cm). Only the back and wing feathers show some greenish-blue coloration, but small size and yellow-orange legs are keys to identification. Found singly at the water's edge in almost any wetland habitat, where it hunts from a low perch or the bank, waiting patiently for prey to venture within striking distance. Common and widespread; to 5,900 ft (1,800 m). Resident population joined by migrants from Sept to April. Range: S Canada to Panama.

Agami Heron
Agamia agami

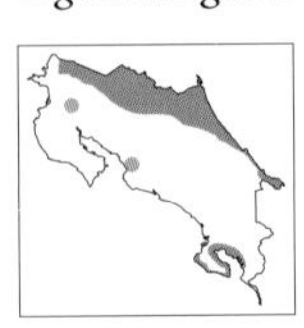

29 in (74 cm). A stunningly beautiful heron, though rarely seen. Its colors allow it to blend in well with the shadowy swamp forest habitat that it prefers. Juvenile is grayish brown dorsally, has brown streaking on flanks, and shares the characteristically long bill. Feeds alone, standing quietly in shallow water or on bank, or by walking very stealthily. Rare in Caribbean lowlands and in Golfo Dulce-Osa Peninsula region; has also been reported occasionally from Lomas de Barbudal, and there is one report from the Tárcoles mangroves (Jan 2007). Range: S Mexico to N Bolivia.

adult

Black-crowned Night-Heron
Nycticorax nycticorax

juvenile

25 in (64 cm). The coloration and habits of adult are somewhat similar to those of the Boat-billed Heron (p. 43), but the bill is that of a typical heron. Juvenile closely resembles juvenile Yellow-crowned Night-Heron (p. 42), but has a mostly pale lower mandible. Most often seen when flying at dawn or dusk, or while roosting by day in trees overhanging water. Fairly common in lowlands around Gulf of Nicoya, rare in Caribbean lowlands and Central Valley; to 3,300 ft (1,000 m). Range: Cosmopolitan.

Yellow-crowned Night-Heron
Nyctanassa violacea

adult

24 in (61 cm). Though the crown is actually cream-colored, the adult is easily recognized by the bold head pattern. Juvenile closely resembles juvenile Black-crowned Night-Heron (p. 41), but has an entirely dark bill. Despite its nocturnal habits, it is often seen during the day loafing on sandbars and salt pond dikes, or roosting in trees, especially mangroves. Fairly common and widespread in lowlands; resident population joined by migrants from Oct to April. Range: US to Peru and Brazil.

juvenile

Boat-billed Heron
Cochlearius cochlearius

adult

juvenile

20 in (51 cm). The extremely wide bill makes it unmistakable—and also resulted in this nocturnal species being placed in its own family at one time. Apparently, the odd bill is not an evolutionary adaptation for exploiting a special feeding niche, but rather a part of sexual display, since stomach contents have shown the diet is similar to that of the two night-herons. Loud beak-snapping noises are part of its display ritual. The local common name *chocuaco* derives from the deep, throaty cackling *ka, ka, ka, cua, cua, ho, ho, cuah* sounds produced when disturbed at day roosts. Colonies nest and roost by day in vegetation that overhangs ponds and streams. Fairly common and widespread in lowlands; to 2,000 ft (600 m). Range: Mexico to NE Argentina.

Ibises, Spoonbills (THRESKIORNITHIDAE). The members of this cosmopolitan family come in two flavors: ibises and spoonbills. The ibises have long, decurved bills that taper to a point. The spoonbills are characterized by long, straight bills that flatten and widen towards the tip. Occurring from sea level to as high as 16,400 ft (5,000 m), most species are associated with wetland habitats, where they feed primarily on aquatic invertebrates, as well as small fish and amphibians; a few species prefer drier sites and tend to consume mostly arthropods and small reptiles. All fly with the neck and legs extended. World: 33, CR: 5

White Ibis
Eudocimus albus

adult

25 in (64 cm). No other white wading bird in CR has red bill and legs. Juvenile could be confused with a Whimbrel (p. 53), but has white belly and rump and lacks head stripes. The long bill is used to probe into the soft substrate of mudflats, marshes, and streams. Common in Caño Negro region and around Gulf of Nicoya, uncommon elsewhere in lowlands and in western Central Valley; to 3,000 ft (900 m). Range: S US to SE Brazil.

juvenile

Green Ibis
Mesembrinibis cayennensis

21 in (53 cm). Unlike other local members of the family, this species is somewhat retiring and prefers forested areas, although it is sometimes found at the edge of more open marsh habitat, as in Caño Negro. Individuals or pairs probe the muddy bottoms of swamps and streams. When perched in a tree, can resemble a Black Vulture (p. 58)—until the bill is seen! At dawn and dusk, gives a hollow, rolling *krwa-krwa-krwa-krwa*. Uncommon in Caribbean lowlands; occasionally to 4,600 ft (1,400 m). Range: SE Honduras to NE Argentina.

Roseate Spoonbill
Platalea ajaja

32 in (81 cm). The only pink wading bird in CR. While foraging, the uniquely shaped bill is swept from side to side as the bird stands or wades in shallow water. When a potential prey item is felt, the beak snaps shut. Roosts and breeds communally. Fairly common in Caño Negro region and around Gulf of Nicoya, uncommon elsewhere in lowlands and in western Central Valley; to 3,000 ft (900 m). Range: S US to N Argentina.

Storks (CICONIIDAE). Storks are large wading birds found in the tropics and temperate zones. Temperate zone species migrate to warmer climates in winter. Even resident tropical species regularly fly (mostly by soaring with legs and neck outstretched) sizable distances to food sources. Diets vary with species, though all are carnivorous. Prey items include fish, amphipians, reptiles, small mammals, insects, snails, and, in two Old World species, carrion. Nests are typically built of sticks and contain from three to five eggs. Young of smaller species may fledge within 50 days of hatching, while larger species can take up to 100 or more days. World: 19, CR: 3

Jabiru

Jabiru mycteria

adult

52 in (132 cm). One of the largest birds in CR, the impressive Jabiru is set apart from other white birds by its enormous bill, which is slightly upturned. Individuals or pairs wade in shallow freshwater, where they stab at fish (often eels). A pair may reuse a nest—which can thus become a huge mass of sticks and mud—built in a large tree. Uncommon in Palo Verde and Caño Negro regions. Range: S Mexico to NE Argentina.

adult in flight

Wood Stork
Mycteria americana

adult in flight

adult

40 in (102 cm). Told apart from other white wading birds by the heavy, slightly decurved bill, which is swept from side to side when foraging in shallow, often murky, water. A tactile sense allows the bill to snap shut on prey instantly when contact is made. Feeds primarily on small fish, but also known to take hatchling sea turtles as they emerge onto the sand. Nests colonially in trees overhanging water. Common in Caño Negro region and around Gulf of Nicoya, uncommon elsewhere in lowlands and in western Central Valley; to 3,000 ft (900 m). Range: S US to N Argentina.

Limpkin (ARAMIDAE). The Limpkin, placed in a family by itself, is adapted to a diet of apple snails (Pomacea) and freshwater mussels. The sharp-tipped bill is deftly employed to extract the organisms from their shells. Though not migratory, birds will readily move from favored habitats if flooding or drought affects their ability to find food. Nests can be constructed on mats of floating vegetation, on the ground, in low tangles of vegetation, or on limbs of trees. In all cases, whatever vegetative material that is nearby is pulled together to form the nest. World: 1, CR: 1

Limpkin
Aramus guarauna

26 in (66 cm). Streaked upperparts somewhat resemble those of juvenile night-herons (pp. 41-42), but the bill is longer and slightly decurved. Juvenile White Ibis (p. 44) has a more decurved bill and no streaking. Generally solitary, though numerous in proper conditions; forages in freshwater marshes; perches on low limbs. Gives a loud, wailing *krrAAOOoow*. Fairly common only in Palo Verde NP and Caño Negro regions; uncommon elsewhere in suitable habitat, including the Central Valley; to 4,900 ft (1,500 m). Range: SE US to N Argentina.

Thick-knees (BURHINIDAE). Mostly subtropical and tropical in distribution, one species breeds in Europe and central Asia and migrates south after breeding; all other species are sedentary. Birds of open, usually dry, habitats, thick-knees are active from dusk to dawn, and their large, yellow eyes suggest such nocturnal habits. They feed primarily on invertebrates, but opportunistically catch small vertebrates as well. The nest is a small scrape on the ground where two eggs are laid. The young are precocial. World: 10, CR: 1

Double-striped Thick-knee
Burhinus bistriatus

19 in (48 cm). The long white superciliary, bordered above by a dark lateral crown stripe, distinguishes it from any of the somewhat similar plovers and sandpipers. Pairs or small groups spend the daylight hours standing or sitting in pastures and cleared fields, where they are easily overlooked due to their cryptic coloration and stillness. If disturbed, they generally prefer to walk away rather than fly. Active and vocal at night, uttering a raucous, lengthy *pip-prri-pippridipip-pip-pippridipip...* . Fairly common in northern Pacific lowlands; uncommon in western Central Valley; and occasional reports from central Pacific lowlands and northern central Caribbean lowlands, where status is uncertain; to 2,600 ft (800 m). Range: S Mexico to N Brazil.

Plovers, Lapwings (CHARADRIIDAE). The lapwings and plovers are characterized by rounded heads and large eyes. Together with the numerous species of the sandpiper family (SCOLOPACIDAE) and a few other smaller families, they make up the group of birds known as shorebirds. While not all members of the family are limited to seaside habitats, they are only found in open areas, including fields (both grassy and tilled), arid plains, and tundra. The nest is a scrape on the ground that is sometimes lined with nearby available material (e.g., pieces of leaves, lichens, shells, or pebbles). The young are precocial. Species breeding at higher latitudes are migratory, with some being among the farthest traveling migrants in the world. World: 67, CR: 10

Southern Lapwing
Vanellus chilensis

adult

adult in flight

14 in (36 cm). This handsomely patterned plover is not likely confused. Pairs or family groups favor open grassy areas near water, including rivers, natural ponds, fish ponds, and sewage treatment plants. Aggressive in behavior and quite vocal; repeats a loud, agitated *KEEEAH, KEEAH, KEEAH…*; several individuals often join in. First recorded in CR in 1997, this South American species has become established in lowlands of both Caribbean and Pacific slopes; to 2,600 ft (800 m). Uncommon, but still spreading on both slopes. Range: Honduras to S Chile and S Argentina.

Black-bellied Plover

Pluvialis squatarola

breeding

11 in (28 cm). Overall grayish appearance helps in indentification, but the diagnostic feature is the black axillar, visible in flight. Birds in breeding plumage (i.e., with black bellies) occasionally seen in Aug and May. Mostly seen on tidal flats of rivers, but also on beaches and in mangroves; occasionally farther inland in migration. Common migrant on Pacific coast and uncommon on Caribbean coast, from Aug to May; some juveniles are summer residents. Range: Cosmopolitan; holarctic breeder that winters south to higher latitudes in southern hemisphere.

nonbreeding

Collared Plover

Charadrius collaris

6 in (15 cm). This small neotropical plover is the only local member of the genus that does not have a white nuchal collar. Individuals or pairs forage among the flotsam above the high-tide line on beaches; also forages on river tidal flats and gravel bars. When foraging, it exhibits the classic plover behavior of remaining motionless while looking for a small arthropod, then running it down with a quick burst of speed. Uncommon on both coasts. Range: Mexico to central Chile and Argentina.

Stilts, Avocets (RECURVIROSTRIDAE). Very long legs and long, thin bills typify the stilts and avocets—the latter with upcurved bills. Cosmopolitan in distribution, they occur in open wetlands, from coastal marshes to high Andean lakes, where they forage using both vision and the tactile sense of their long bills (under water or in mud) to find small invertebrates. Nest is either a mere scrape on dry ground or a construction of sticks and vegetation when built on floating vegetation. Nesting is often colonial; the chicks are precocial. World: 9, CR: 2

Black-necked Stilt

Himantopus mexicanus

15 in (38 cm). Arguably the most striking of CR shorebirds, the black-and-white plumage and exaggeratedly long pinkish legs should preclude any confusion. The erect posture and alert behavior give rise to the local common name of *soldadito* (little soldier). Inhabits marshes, tidal flats, and ponds, including salt ponds; gregarious and often in groups of a dozen or more. These easily excited birds produce long series of sharp *kek* notes. Resident birds are common around Gulf of Nicoya; migrants are likewise common in this area and are fairly common elsewhere in lowlands and in Central Valley to 4,900 ft (1,500 m), from Oct to May. Range: US to S SA.

Sandpipers, Allies (SCOLOPACIDAE). Though this cosmopolitan family comprises the prototypical shorebirds, few species actually breed along coasts, which are the preferred wintering habitat of many migratory species. In fact, an array of environments—including open ocean, temperate forests, and arctic tundra—provide habitat for these birds, whose bodies and bills vary greatly in size and shape. Food items are also diverse, but consist mostly of invertebrates plucked directly from substrates, or, perhaps more commonly, probed for in water or soft soil (employing tactile receptors in the bill tip). Most species are ground nesters, though a few occupy used thrush nests in trees. The precocial chicks forage for themselves but are brooded by at least one adult. World: 96, CR: 32

Spotted Sandpiper

Actitis macularius

breeding

8 in (20 cm). One of the most common and widespread sandpipers in CR, this species can be found in a variety of habitats, including the edges of freshwater ponds, along rocky montane streams, and on coastal mudflats and sandbars, where widely scattered individuals forage for small arthropods. Most of the individuals seen in Costa Rica are in nonbreeding plumage and thus lack the distinctively spotted undersides they sport on their NA breeding grounds. Nonetheless, they are readily identified by the grayish-brown smudge on the side of the breast, faint superciliary and eye line, and nearly constant bobbing motion of the posterior end of the body. In flight, they expose a narrow white wing stripe. Common and widespread migrant from early Aug to mid-May, rare summer resident; to 7,200 ft (2,200 m). Range: Breeds in NA, winters south to N Chile and Argentina.

nonbreeding

Willet
Tringa semipalmata

breeding

nonbreeding

15 in (38 cm). A largish, pale gray, fairly nondescript shorebird in nonbreeding plumage. In all plumages, birds in flight show a broad white stripe across the black wings. Mostly found in brackish wetlands (e.g., mudflats and mangroves), less commonly along seashores; often with Whimbrels. Regularly rests standing on one leg, and even hops one-legged for short distances, as do many shorebirds. In flight, gives a loud *klee-klee-klee*. Migrants are common on Pacific coast and uncommon on Caribbean coast, from Aug to May; uncommon summer resident on Pacific coast. Range: Breeds from central and SE Canada to West Indies, winters south to S Brazil and N Chile.

Whimbrel
Numenius phaeopus

17 in (43 cm). The combination of crown stripes and a long, decurved bill is diagnostic. Uses its odd-looking bill to extract crabs from their burrows in fresh mud. Mostly found in brackish wetlands (e.g., mudflats and mangroves), less commonly along seashores; often with Willets. Migrants are common on Pacific coast and uncommon on Caribbean coast, from Aug to May; fairly common summer resident on Pacific coast. Range: Cosmopolitan; holarctic breeder that winters south to higher latitudes in southern hemisphere.

Ruddy Turnstone
Arenaria interpres

breeding

9 in (23 cm). The most boldly patterned of Costa Rica's regularly occurring sandpipers, even in nonbreeding plumage. Individuals or groups forage on mud and sand exposed by the receding tide, as well as on rocky coasts. Often uses its wedge-shaped bill to flip over small stones, shells, or debris in search of hidden prey. Migrants are common on Pacific coast and uncommon on Caribbean coast, from Aug to late May; uncommon summer resident on both coasts. Range: Cosmopolitan; holarctic breeder that winters south to higher latitudes in southern hemisphere.

nonbreeding

Sanderling
Calidris alba

breeding

8 in (20 cm). This is the classic sandpiper that runs back and forth along the water's edge on sandy beaches. During most of their stay in Costa Rica, these birds can be distinguished by the combination of black bill, eye, and legs, together with white underparts and very pale upperparts that show touches of black and gray. Rufous head and breast are distinctive in breeding plumage. In both plumages, birds in flight show bold white wing stripe. Fairly common migrant from mid-Aug to early May, rare summer resident; occurs along both coasts. In addition to beaches, also found on mudflats and around salt ponds. Range: Cosmopolitan; holarctic breeder that winters coastally worldwide.

nonbreeding

Jacanas (JACANIDAE). The distinctive feature of jacanas is their impressively long toes and claws, which enable them to walk on floating vegetation by efficiently distributing their weight—a behavior that gives rise to their common name *lily-trotter*. The diet comprises mostly small insects plucked from plant or water surfaces, but also aquatic invertebrates, small fish, and some plant matter. Nests are platforms of aquatic plants constructed on floating vegetation. Apart from one African species, the females of this pantropical family are polyandrous and have no domestic duties, leaving their respective males to incubate the eggs. The precocial chicks are led around by the male, but feed themselves. World: 8, CR: 2

Northern Jacana
Jacana spinosa

adult

juvenile

9 in (23 cm). The adult, with its long legs and toes and bright yellow-orange bill and frontal shield, is easily identified. Juvenile has a brownish back and striped head and neck, causing potential confusion with other shorebirds, but in flight shows yellow flight feathers (as does the adult) and holds wings open briefly upon landing. Inhabits marshes, ponds, wet pastures, and riverbanks. Readily utters loud, high twittering notes, often with an agitated quality. Common and widespread; to 4,900 ft (1,500 m). Range: N Mexico to W Panama.

Rails, Crakes, Gallinules (RALLIDAE). This cosmopolitan family includes some of the world's most enigmatic birds, due to their typically skulking habits. With long legs, short tails, and slender bodies, they are adept at slipping through dense vegetation. The preference of many species for marsh habitats makes them hard to see. Though not strong fliers, they have colonized numerous remote islands; they probably arrive by being blown off course and then survive by virtue of their omnivorous diet. Pairs weave open or domed nests in vegetation. Chicks are precocial. World: 141, CR: 16

White-throated Crake
Laterallus albigularis

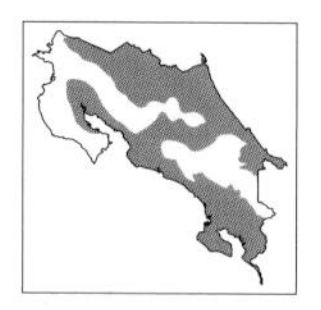

Caribbean race

Pacific race

6 in (15 cm). The combination of a rufous breast and barred posterior underparts is unique among CR members of the family. Inhabits almost any damp grassy area; though very furtive, occasionally reveals itself at edges and openings. Frequently utters a loud, descending *churr*, lasting 5 seconds or more. Common and widespread in wet lowlands and foothills; also common in eastern Central Valley; to 4,900 ft (1,500 m); rare north of Quepos. Range: SE Honduras to W Ecuador.

Gray-necked Wood-Rail
Aramides cajaneus

15 in (38 cm). Somewhat chicken-like in appearance. The yellow-orange bill and pink legs distinguish it from all but the similar Rufous-necked Wood-Rail *Aramides axillaris* (not illustrated) of Pacific mangroves—so check the neck color to be sure. Can be found in most forest habitats, though rarely far from water. The resounding call, most often heard at twilight and bringing to mind a group of drunken chickens, is given in chorus: *káh-klah, káh-klah, kah koh-ho-ho-ho-hah*. Common and widespread in wet lowlands, foothills, and Central Valley; uncommon in northern Pacific; to 4,600 ft (1,400 m). Range: S Mexico to N Argentina.

adult

juvenile

Purple Gallinule
Porphyrio martinicus

13 in (33 cm). The adult's two-toned bill (with yellow tip and red base), which extends up onto forehead as a light blue frontal shield, is distinctive. Juveniles lack the colorful bill and are buffy-brown above and whitish below. Regularly seen at the edges of wetland habitats as it walks along muddy margins or clambers through reeds and bushes. Gives a high-pitched *kick*; also cackles like a hen. Fairly common and widespread; to 4,900 ft (1,500 m). Range: S US to N Argentina.

New World Vultures (CATHARTIDAE). The New World vultures are highly adapted for their specialized diet of carrion and, as such, are important components of natural ecosystems. Despite having strong, sharp beaks, vultures seldom kill their own food. Instead, they rely on being able to soar over large distances in search of recently dead animals. Three species have evolved an excellent sense of smell, while the other four use vision to find their food. Many species roost communally, but all are solitary nesters. One or two eggs are laid directly on the ground under dense shrubbery, on the floor of a rock crevice, or in a hollow log. World: 7, CR: 4

Black Vulture

Coragyps atratus

adult

25 in (64 cm). Countrywide, it is generally the most commonly seen soaring bird below timberline, though it is less common in extensively forested areas. Easily mistaken for a hawk, the unfeathered black head and white patch near the wing tip are diagnostic features. Typically soars with wings held flat. Uses vision—and follows Turkey Vultures—to find food. Will kill small animals such as hatchling sea turtles; also visits garbage dumps, and will even eat fruit (e.g., oil palm and bananas). Range: US to Argentina.

Adult with white wing patches exposed.

Turkey Vulture
Cathartes aura

adult

adult in flight

30 in (76 cm). The unfeathered red head and pale gray flight feathers distinguish it from other vultures and birds of prey. When soaring, holds wings angled up in a shallow V. Unlike most birds, the members of this genus have acute olfaction, enabling them to zero in on small carcasses hidden beneath the forest canopy. Common resident countrywide, though uncommon above 6,600 ft (2,000 m). Spectacular numbers of passage migrants traverse CR from mid-Sept to late Oct and from late Jan to early May, mostly in Caribbean lowlands. Range: S Canada to Tierra del Fuego.

King Vulture
Sarcoramphus papa

32 in (81 cm). The multi-hued head is distinctive, though birds often are seen soaring so high that the colors are not easily discerned, and identification is then based on white body with black tail and flight feathers. (The similarly patterned Wood Stork [p. 47] has long bill and legs.) Despite being the largest vulture in CR and dominant at carcasses, it is decidedly uncommon in lowlands and foothills; to 3,900 ft (1,200 m). Prefers areas where mature forest still exists, though readily soars over open country. Range: S Mexico to N Argentina.

Osprey (PANDIONIDAE). Closely related to the Accipitridae, the Osprey is currently considered distinct enough to merit being in its own monotypic family. Its specialized diet of fish has driven the evolution of unique characteristics, including spiny pads on the soles of its feet and a reversible outer toe, both of which aid in holding on to its slippery prey. Likewise, it has valves that seal the nostrils when diving into water and oily plumage that affords waterproofing. Pairs construct very large nests of sticks and finer material placed atop trees and, nowadays, manmade structures such as utility poles. Hit hard by DDT contamination in the 1950s and 1960s, populations have since recovered. World: 1, CR: 1

Osprey
Pandion haliaetus

adult

adult in flight

23 in (58 cm). White crown and underparts are separated by a dark eye line. Seen perched on high, exposed branches or flying with distinctively crooked wings. Almost always near water, where it hunts for fish, typically by hovering, then plunging feet first into the water. Common winter resident from early Sept to April, to 9,200 ft (2,800 m); common passage migrant along coasts from early Sept to Oct and from March to April; uncommon summer resident from May to Aug, when it is known to occur at Lake Arenal and in southern Pacific lowlands (and likely elsewhere). Range: Cosmopolitan.

Hawks, Kites, Eagles (ACCIPITRIDAE). The members of this large, cosmopolitan family, which includes the Old World vultures, share characteristics such as a cere (often colorful, and helpful in ID), sharp-edged beak with a hooked tip, sharp talons, and keen vision. Beyond that, they have diversified greatly in order to exploit a wide range of habitats and food items (from snails to monkeys). In CR, they range in size from the 8 in (20 cm) male Tiny Hawk *Accipiter superciliosus* (not illustrated) to the 42 in (107 cm) female Harpy Eagle *Harpia harpyja* (not illustrated). A pair builds a nest of sticks, lined with leaves. Juveniles of most species differ from adults in plumage. World: 243, CR: 39

White-tailed Kite
Elanus leucurus

adult, dorsal view

adult, ventral view

16 in (41 cm). An attractive hawk; the black shoulders and white tail are diagnostic. Flies leisurely and often hovers while hunting; occurs in open areas with scattered trees. Most prey is taken from the ground and consists of rodents, lizards, and insects. Since its arrival in CR in the mid-1950s, this species has now become fairly common countrywide; to 4,900 ft (1,500 m). Range: US to central Chile and Argentina.

Swallow-tailed Kite
Elanoides forficatus

adult, ventral view

adult, dorsal view

23 in (58 cm). Almost always seen in flight, when the long wings and forked tail, combined with white-and-black coloration, make it easy to ID. A graceful aerialist, it circles, glides, and swoops over treetops in search of large insects and small lizards. Consumes prey while in flight. Common breeding migrant from late Dec to mid-Sept, mostly in foothills and mountains; generally absent from western Central Valley and northern Pacific; some year-round residents occur in southern Pacific and on Caribbean slope. Range: Breeds from SE US to Panama, winters south to NE Argentina.

Gray-headed Kite
Leptodon cayanensis

20 in (51 cm). Adult is only CR raptor that has gray head, black back, and white underparts; also note bluish-gray soft parts. In flight, recalls a hawk-eagle, but combination of black wing linings and white body is unique. Juveniles are variable in plumage with pale, intermediate, and dark morphs—all have yellow-orange soft parts. Perches in canopy and at forest edges; soars on broad wings. Fairly uncommon in lowlands and foothills; to 3,000 ft (900 m). Range: SE Mexico to N Argentina.

Ornate Hawk-Eagle
Spizaetus ornatus

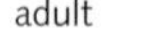
adult

juvenile

24 in (61 cm). A large hawk with dramatic plumage. Juvenile combines mostly white head and breast with barred thighs. Perches inconspicuously at all levels inside mature wet forest, where it hunts for medium-sized vertebrates. Most often seen soaring during sunny mornings, when it draws attention to itself with loud whistles: *WEE-EW, WEEH, WEH-WEH-WEH.* Uncommon in wet lowlands, foothills, and middle elevations of Caribbean and southern Pacific slopes; to 7,200 ft (2,200 m). Range: S Mexico to N Argentina.

Double-toothed Kite
Harpagus bidentatus

male

14 in (36 cm). A small raptor that is potentially hard to ID, but on perched birds note the dark stripe on center of white throat; on flying birds, note the puffy white undertail coverts. Regularly soars on sunny mornings; perches at middle levels of forest, second growth, forest edges, and gardens; often follows foraging troops of White-throated Capuchins or Central American Squirrel Monkeys in order to catch the lizards and large insects that the monkeys flush out. Fairly common; to 5,600 ft (1,700 m). Range: SE Mexico to SE Brazil.

female

juvenile

Bicolored Hawk
Accipiter bicolor

adult

Male 14 in (36 cm); female 17 in (43 cm). Two-toned gray adult has diagnostic rufous thighs. Juveniles are browner above, vary from white to buffy below, and do not always have rufous thighs. Ambushes avian prey from perches at almost any level in mature wet forest, tall second growth, forest edge, and gardens. Gives a series of more than a dozen barking, trogon-like notes: *keh-keh-keh-keh...* . Rare in humid and wet regions; to 5,900 ft (1,800 m). Range: S Mexico to Tierra del Fuego.

juvenile

Semiplumbeous Hawk
Leucopternis semiplumbeus

15 in (38 cm). The combination of dark gray upperparts and white underparts, together with orange cere and legs, sets it apart. A rather inconspicous raptor, unless calling; call typically consists of a short series of upslurred notes: *klerEE-klerEE-klerEE*. Otherwise, perches quietly at middle levels of mature wet forest, advanced second growth, and forest edges; does not soar. Mostly takes small vertebrates. Fairly common in Caribbean lowlands, rarely above 1,600 ft (500 m). Range: E Honduras to NW Ecuador.

White Hawk
Pseudastur albicollis

adult

adult in flight

23 in (58 cm). A most handsome hawk, whether perched or flying; the only CR raptor with an all-white back. Soars regularly on broad wings. Perches quietly at middle levels of mature wet forest, advanced second growth, and forest edges; occasionally follows foraging troops of monkeys in order to catch prey that they stir up. Seems to specialize on snakes and lizards, but will take other small vertebrates and large insects. Fairly common in Caribbean lowlands and foothills, on Pacific slope of Guanacaste Cordillera, and from Carara NP south; rare above 3,900 ft (1,200 m); also rare on southern Nicoya Peninsula. Range: SE Mexico to Amazonian Brazil.

Common Black Hawk
Buteogallus anthracinus

adult

juvenile

22 in (56 cm). The adult's yellow lores set it apart from other black raptors. In flight, the white tail band eliminates confusion with Black Vulture (p. 58). Juvenile resembles juveniles of several other hawk species, but note barred undertail coverts. Feeds primarily on crabs; perches at lower and middle levels, and often quite confiding. Common along Caribbean and Pacific coasts, almost always near water, especially in and around mangrove swamps and along beaches; rare elsewhere in lowlands. Resident on Caño Island. Range: SW US to N SA and N Peru.

Roadside Hawk

Rupornis magnirostris

adult

15 in (38 cm). One of the more commonly seen hawks in CR, due largely to its preference for open areas with scattered trees and its habit of perching fairly low, even on fence posts. Though it seldom soars, a diagnostic feature is the rufous color on the primaries, visible in flight. Adult can be told apart from the somewhat similar adult Gray Hawk by the rufous barring on the breast. Juvenile has rufous barring on thighs. Opportunistically takes large insects and small vertebrates. Widespread and fairly common; to 4,900 ft (1,500 m). Range: Mexico to N Argentina.

adult in flight

Broad-winged Hawk
Buteo platypterus

adult

juvenile

16 in (41 cm). A participant in one of the most impressive avian spectacles in CR, when hundreds to thousands of migrating birds (including Turkey Vultures [p. 59] and the somewhat larger and longer-winged Swainson's Hawk *B. swainsoni* [not illustrated]) traverse the country from Oct to mid-Nov and again from early March to May. The dark malar stripe helps in identifying both adult and juvenile; also note adult's tail bands, visible in flight. Perches at middle levels of forest edges; soars frequently. Abundant passage migrant; to 11,500 ft (3,500 m). Common and widespread winter resident from Sept to May, mostly below 6,600 ft (2,000 m). Range: Breeds from Canada to S US, winters south to Bolivia.

Gray Hawk
Buteo plagiatus

adult

juvenile

17 in (43 cm). The gray barring on the underparts and all-gray wings set it apart from the somewhat similar Roadside and Broad-winged Hawks. Juvenile has whitish superciliary and cheeks separated by dark eye line. Very similar habitats as those of the Roadside Hawk, preferring forest edges and open areas with scattered trees; uses exposed perches, including utility poles and power lines; soars regularly. Fairly common throughout mapped range, south to about Dominical; to 3,600 ft (1,100 m). Range: SW US to CR. (The Gray-lined Hawk *B. nitidus* [not illustrated] of S Pacific lowlands was formerly considered part of this species. Adults have fine gray barring on upperparts and juveniles have more blotchy underparts and, at best, a very indistinct malar stripe.)

Red-tailed Hawk

Buteo jamaicensis

juvenile, dorsal view

22 in (56 cm). Along with the Swallow-tailed Kite (p. 61), the resident race is one of the two most common highland raptors. Both residents and NA migrants are readily distinguished from other raptors by the tail color. Resident race freely soars and often hangs stationary in the wind; prefers partially cleared areas in montane regions and high, exposed perches. Screams a high, grating *keEEaah*. Resident race (indicated on map) is fairly common in highlands above 4,900 ft (1,500 m), and uncommon from 4,900 ft (1,500 m) down to about 2,600 ft (800 m). Migrants (not indicated on map) are uncommon from Oct to April and can turn up almost anywhere, even in lowlands. Range: NA to Panama.

adult, ventral view

Falcons, Caracaras (FALCONIDAE). Long considered to be close relatives of hawks, recent DNA evidence has shown that falcons are in fact an example of convergent evolution, in which both families independently evolved similar structures (e.g., sharp beak and strong talons) for a predatory existence. The members of this cosmopolitan family inhabit almost every imaginable habitat from desert to rainforest and have evolved several discrete foraging strategies. Only the caracaras construct stick nests; other species lay eggs in tree hollows, on cliff ledges, directly on the ground, and sometimes on manmade structures. Most species are solitary pair nesters, with the male doing most of the provisioning while the female does most of the incubation and brooding. World: 65, CR: 13

adult

Crested Caracara

Caracara cheriway

juvenile

23 in (58 cm). The adult is quite distinctive, with its black-and-white plumage and bare red-orange cere and facial skin. The juvenile is a muted version of the adult in brown and buff, and can be told apart from the Yellow-headed Caracara (p. 70) and Laughing Falcon (p. 71) by its dark crown and belly. Prefers agricultural areas; perches high in trees as well as on the ground; feeds primarily on carrion, but also attacks and kills small vertebrates. Common in northern Pacific lowlands and foothills; with deforestation, it's becoming increasingly common in southern Pacific lowlands and foothills, and in northern central Caribbean lowlands; uncommon in Central Valley to east of Cartago; to 4,900 ft (1,500 m). Range: S US to N Peru and Brazil.

Yellow-headed Caracara
Milvago chimachima

16 in (41 cm). In coloration, somewhat similar to the Laughing Falcon, but shows only a thin dark line behind the eye. Juvenile has streaking on head and underparts. Found in agricultural areas; perches high in trees as well as on the ground, and even on the backs of cattle; feeds mostly on carrion. Utters a dreadful, nasal scream: *reeee-aaah!* Since first CR record in 1973, has become common in southern and central Pacific and in the western Central Valley; uncommon north of Gulf of Nicoya on the Pacific slope, in the Caño Negro region, and in the eastern Central Valley from Cartago to Turrialba; to 4,900 ft (1,500 m). Range: CR to N Argentina.

adult

juvenile

adult in flight

Laughing Falcon

Herpetotheres cachinnans

20 in (51 cm). The broad dark mask that makes a complete ring behind the head is the distinctive field mark of this attractive raptor. Often perches conspicuously at forest edges and in open areas with scattered trees; feeds primarily on snakes. Calls mainly at dawn and dusk with a loud, far-carrying *GWA-CO* (the first note higher and louder), thus its local common name: *guaco*. The full song is often given in duet, starting with single *GWA!* notes and accelerating into a crazed laughing cacophony. Fairly common and widespread in lowlands and foothills; generally to 3,900 ft (1,200 m), rarely up to 5,900 ft (1,800 m). Range: N Mexico to N Argentina.

Bat Falcon

Falco rufigularis

Male 10 in (25 cm); female 12 in (31 cm). In flight, could be mistaken for the similarly sized and patterned White-collared Swift (p. 80), though swifts usually travel in flocks. Some birds have orange-rufous throat and breast. Found on high, exposed perches in partly to mostly cleared areas; captures fast-flying prey (swallows, swifts, hummingbirds, and bats) on the wing. The call is a shrill *ki-ki-ki-ki…* . Uncommon in wet lowlands and middle elevations, rare in northern Pacific; generally to 4,900 ft (1,500 m) and rarely as high as 9,500 ft (2,900 m). Range: N Mexico to N Argentina.

Owls (STRIGIDAE). Despite their mostly nocturnal habits, owls are a readily recognized group of birds, owing to the large, rounded head and big-eyed appearance. Almost every terrestrial habitat is occupied by at least one member of this cosmopolitan family. Fearsome predators, they use both vision and hearing to detect prey, which is grasped in powerful, sharp talons. Most species are monogamous. None builds its own nest and eggs are laid in a variety of sites, including tree cavities, the former nests of other birds, cliff ledges, and directly on the ground. Females handle all incubation and brooding, while the male hunts and provides food for his mate and offspring. Vocalizations are important both for locating and identifying owls. World: 207, CR: 16

Pacific Screech-Owl

Megascops cooperi

9 in (23 cm). Four species of screech-owl can be found in CR. They tend to separate out by habitat and can be told apart by voice. This is the only species that occurs in dry forest habitat. Hunts for large insects from low perches in woods, gardens, and open areas. Gives a short *hup* in a fast-paced series that rises then falls abruptly in pitch and tempo. Fairly common in northern Pacific, to 3,300 ft (1,000 m); uncommon in western Central Valley and rare east to Cartago; also reported in Caño Negro region. Range: S Mexico to NW CR.

Tropical Screech-Owl

Megascops choliba

9 in (23 cm). The distinct black border of the facial disk is the best field mark on this small owl. Hunts from low perches in trees, in both gardens and lightly wooded areas. Gives a bubbly, one-second trill followed by a slightly louder, questioning *POO!?* (this last note is sometimes given two or three times). Fairly common in Central Valley and intermontane valleys of southern Pacific, from 1,600 to 4,900 ft (500 to 1,500 m); uncommon to rare in Tilarán Cordillera (to 4,300 ft [1,300 m]) and in Pacific lowlands south of Orotina. Range: CR to N Argentina.

Crested Owl
Lophostrix cristata

16 in (41 cm). The white eyebrows extending to long, white ear tufts impart an imposing aspect to this medium-sized owl. Hunts mostly for large insects. Roosts at low to middle levels of forest and adjacent tall second growth; also at forest edges. Call is a deep, guttural *wk-wkwuUUuurr* that could be mistaken for a distant Bare-throated Tiger-Heron growl (p. 34). Widespread but generally uncommon; to 4,900 ft (1,500 m). Range: S Mexico to N Bolivia.

adult unbarred morph

Spectacled Owl
Pulsatrix perspicillata

19 in (48 cm). The largest owl likely to be seen in CR (Great Horned Owl *Bubo virginianus* [not illustrated] is extremely rare). Has a more sharply contrasting facial pattern than the Mottled Owl (p. 74); note plain buffy underparts, though some birds in NW Pacific show some barring. Roosts fairly low in trees, typically near wooded streams. Hunts for small to medium-sized mammals, as well as insects, lizards, and occasional birds from perches in mid-canopy. Call is a series of about six deep, muffled, chuckling notes that sounds like a large sheet of tin being shaken. Fairly common and widespread; to 4,900 ft (1,500 m). Range: S Mexico to N Argentina.

adult barred morph

juvenile

Ferruginous Pygmy-Owl
Glaucidium brasilianum

adult, dorsal view

adult, ventral view

6 in (15 cm). The most often seen of the three pygmy-owls in CR and the only one in its range. A fierce predator, it takes insects, lizards, and even small birds. Active day and night at all levels of dry and humid forest, forest edges, and gardens. It is often mobbed by other birds and has false eye spots on the nape to help avoid being struck from behind. Gives a monotone series of rather staccato toots. Common in lowlands and foothills of northern Pacific and western Central Valley, rarer east of San José to Turrialba and south of Jacó to Dominical; to 7,200 ft (2,200 m). Range: SE Arizona to central Argentina.

Mottled Owl
Ciccaba virgata

13 in (33 cm). This dark brown owl can be distinguished by its streaked belly and brown iris. Forages for large insects and small mammals at middle levels of forests, tall second growth, and forest edges; roosts fairly low in dense vegetation. Calls with deep, muffled notes followed by two loud notes and an additional muffled note: *whu, whu, WHUU, WHUU, whu.* Fairly common and widespread; to 7,200 ft (2,200 m). Range: N Mexico to N Argentina.

Black-and-white Owl
Ciccaba nigrolineata

15 in (38 cm). No other local owl has completely barred underparts. Forages at forest edges and in gardens, often near street lamps, where it catches large insects, as well as the bats that come to eat them. Roosts in lower and middle levels of large trees. Gives an ascending series of deep chuckling notes: *hwa,hwa, hwa, hwa, hwa, HWUU!* Fairly common in wet lowlands, rare in dry forest and middle elevations; to 4,900 ft (1,500 m). Range: S Mexico to NW Peru.

Striped Owl
Pseudoscops clamator

15 in (38 cm). The obvious ear tufts, black-bordered facial disk, and heavily streaked underparts are distinctive. Favors open areas near woods, where its predominant prey consists of small mammals; often seen perched on utility wires. Nests and roosts on or near ground. Makes a nasal, high-pitched *heeAHhh* that rises and falls. Fairly common in central Pacific lowlands, especially near oil palm plantations; uncommon elsewhere in mapped range: to 4,900 ft (1,500 m). Range: S Mexico to N Argentina.

Potoos (NYCTIBIIDAE). This neotropical family is strictly nocturnal in its foraging habits. The diet comprises flying insects, along with the occasional small bat, swallowed up in flight by the bird's huge gape. Unlike their relatives the nightjars, potoos roost and hunt in an upright position. Roost sites are often in exposed situations, but the birds' cryptic plumage allows them to blend in as if part of the branch or stub. A single egg is laid (rather precariously) in a depression on a limb or at the end of a vertical snag. Both parents attend to nesting duties and the young fledges nearly three months after egg-laying. World: 7, CR: 3

Great Potoo
Nyctibius grandis

20 in (51 cm). The largest of the potoos, it can be told apart from other local species by the lack of a dark malar stripe. During the day, it typically perches crosswise on a high limb inside mature forest or at forest edges. At night, it sallies for flying insects from high, exposed perches. Gives a loud, throaty *WAAAAuur* that falls off slightly at the end and also a quick, hollow *WAU!*; most often heard on moonlit nights. Uncommon in wet lowlands of Caribbean slope as well as Osa Peninsula and Golfo Dulce area; to 2,000 ft (600 m). Range: S Mexico to SE Brazil.

Common Potoo
Nyctibius griseus

15 in (38 cm). Virtually identical to the Northern Potoo (no known range overlap) and only separable by voice. Often perches quite low, at the tip of a more or less vertical snag (or even fence post), appearing amazingly like an extension of the perch itself; seems to prefer more open areas than Great Potoo. The call is a sweet and mournful, resonant, descending *KWUU-UU, KUU, KUU, kwuu, kwuu*. Fairly common in lowlands and foothills of southern Pacific (though rare in the Carara NP area), uncommon in Caribbean lowlands and foothills; to 3,900 ft (1,200 m). Range: Nicaragua to N Argentina.

Northern Potoo

Nyctibius jamaicensis

15 in (38 cm). Virtually identical to Common Potoo in appearance and habits—separable only by voice. Combines the gruff, throaty sound of a Great Potoo with the cadence of the song of a Common Potoo. If no vocalization is heard, identification can be based on geographic location, but caution is urged where their ranges are in close proximity. Uncommon in northern Pacific lowlands and foothills; to 3,300 ft (1,000 m). Range: Mexico to CR.

Oilbird (STEATORNITHIDAE). The only nocturnal bird in the world that feeds solely on fruit, the Oilbird is a true oddity. Most of its food consists of species in the avocado and palm families that are rich in proteins and fat, and are plucked in hovering flight. A highly developed sense of smell may aid in finding ripe fruits in the dark. Birds roost and nest colonially on ledges in caves and dark ravines, where echolocation assists them in navigation. World: 1, CR: 1

Oilbird

Steatornis caripensis

18 in (46 cm). The reddish-brown plumage is puncutated with white spots on the wings and body. Still a mystery; no CR breeding population has yet been discovered, but regular occurrence in Monteverde (since at least 2009) between June and Sept—a time when South American populations should be nesting—hints at the possibility of a resident population. Additionally, individual records come from Cerro de la Muerte, Corcovado NP, Bijagua de Upala, and above Horquetas de Sarapiquí. Range: CR to N Bolivia.

Nightjars, Allies (CAPRIMULGIDAE). The nighthawks and nightjars are specialists at capturing nocturnal flying insects. They have large eyes to help spot their quarry in the dimness and are capable of opening their mouths extremely wide to swallow up their prey. Most active crepuscularly and when there is bright moonlight, it is often difficult to see these birds well—and indeed they are all quite similar in appearance—so their vocalizations are most helpful in identification. During the day, their cryptically colored plumage helps them avoid detection as they roost in leaf litter, on sandy soil, or on tree limbs. Though the majority of species are tropical, the family occurs throughout most of the world, with temperate zone breeders being migratory. Eggs are laid directly on the ground. World: 97, CR: 10

Lesser Nighthawk
Chordeiles acutipennis

9 in (23 cm). Look for this nighthawk in the last light of day, just after sunset, typically flying fairly high and somewhat erratically as it pursues aerial insects. Forages over open areas, often near water. Roosts lengthwise on low branches, especially in mangroves. Fairly common resident on Pacific slope, especially near coast. Abundant passage migrant, mostly along Caribbean coast, from mid-Sept to early Nov. Range: SW US to Brazil and N Chile; northern populations winter south to S CA and N Colombia.

Common Pauraque

Nyctidromus albicollis

11 in (28 cm). This is the species in this family you are most likely to encounter in CR; seen from dusk to dawn foraging for insects by flying up from the ground in openings, especially roadsides and lawns; usually near woods, where it roosts in leaf litter. If disturbed during the day, it will fly low for a short distance and drop to the ground, blending in with the mottled browns of the fallen leaves. Whistles a slurred *weeeuu!* often introduced by a stuttering series of *wit* notes. Common in lowlands and middle elevations; to 5,600 ft (1,700 m). Range: S Texas to NE Argentina.

Dusky Nightjar

Antrostomus saturatus

9 in (23 cm). No other member of the family is likely at the elevations this species inhabits; nonetheless, best told apart by voice. Hunts for flying insects from low perches at montane forest edges and in clearings; also prefers low perches for roosting. After dark, males whistle a somewhat slurred *pt whEE-per-whEE*, the last note slightly higher pitched. Fairly common on Central and Talamanca Cordilleras, from 5,900 ft (1,800 m) to just above timberline; uncommon on Tilarán Cordillera, above 4,900 ft (1,500 m). Range: CR and W Panama.

Swifts (APODIDAE). Swifts prey exclusively on airborne insects swallowed in flight using the birds' wide gape. Together with the hummingbirds, swifts are the only birds in the world capable of obtaining power from the upstroke of the wing (although they cannot hover as hummingbirds do). Due to a similar aerodynamic design for an aerial lifestyle, swifts somewhat resemble swallows, but can be distinguished by their stiff wingbeats. Nest construction and placement varies in this essentially cosmopolitan family, though numerous species use saliva to help hold the nest together and many nest in inaccessible places such as cliffs, caves, hollow trees, chimneys, and behind waterfalls—they are adept at clinging to vertical surfaces, but incapable of walking on their weak, short legs. World: 104, CR: 11

White-collared Swift

Streptoprocne zonaris

9 in (23 cm). Many swifts pose difficult ID problems. Fortunately, this one—the most widespread and commonly seen species in CR—is fairly straightforward. The white collar is easy to see under all but the poorest lighting conditions, and the large size is distinctive (though a lone individual could be confused with a Bat Falcon [p. 71]). Usually seen traveling and foraging in groups of a dozen or more (up to 100+); commonly with other, smaller swift species. Widespread and generally common, though uncommon in northern Pacific. Range: S Mexico to N Argentina.

Swallows (HIRUNDINIDAE). Placed here next to swifts—for ease of comparison—the swallows are actually completely unrelated birds that have independently evolved the habit of feeding on flying insects. Given that this resource is seasonal at higher latitudes, nine out of the 13 species recorded in CR are migrants. Members of this cosmopolitan family nest in various manners. Some species excavate burrows in earthen banks, while others use burrows made by other birds (e.g., motmots and kingfishers); some species use some vegetation to line existing cavities in trees, rock crevices, and manmade structures, and still others construct a nest of mud or clay. World: 86, CR: 13

Blue-and-white Swallow

Pygochelidon cyanoleuca

4 in (10 cm). This is the common resident swallow of the Central Valley and middle and upper elevations. Dark undertail coverts distinguishes it from all other swallows with white underparts. Small groups forage over open areas, in both urban and rural settings; often fairly high. Common in middle elevations and highlands, from 1,600 to 10,200 ft (500 to 3,100 m); occasionally down to sea level, as at Dominical. Range: CR to Tierra del Fuego.

Gray-breasted Martin

Progne chalybea

7 in (18 cm). The largest of the resident swallows, it has dark bluish upperparts and grayish throat and breast; also note notched tail. Forages over open areas, often near water; roosts and nests in tree cavities, as well as under bridges and in open-ended iron beams of buildings (e.g., gas and bus stations). Fairly common and widespread in both town and country; to 5,600 ft (1,700 m). Range: Mexico to N Argentina.

Mangrove Swallow
Tachycineta albilinea

adult

5 in (13 cm). The combination of blue-green upperparts and white rump and underparts is diagnostic. Juvenile has gray upperparts. Forages by skimming low over water surfaces and fields, often far from mangroves. Common in lowlands, uncommon at higher elevations; to 3,300 ft (1,000 m). Range: N Mexico to Panama.

juvenile

Barn Swallow
Hirundo rustica

5 in (13 cm). The long, pointed wings and tail impart a sleek look; juvenile has shorter, less deeply forked tail. On adult, lower underparts are pale rufous. Juvenile has buffy-white underparts. In migration, occurs virtually everywhere, but with largest numbers flying low along coasts (often with Cliff Swallows *Petrochelidon pyrrhonota* and Bank Swallows *Riparia riparia* [not illustrated]); winter residents prefer open fields. When not flying, often perches on wires and fence lines. Abundant and widespread migrant from early Aug to early June; rarely to 9,800 ft (3,000 m). Range: Cosmopolitan.

Hummingbirds (TROCHILIDAE). After the flycatchers and the recently revised tanagers, this strictly New World family is the third most species diverse in the hemisphere. Hummingbirds have a number of evolutionary adaptations for feeding on flower nectar, the most notable of which is their unique ability to hover and fly backward, indeed in any direction. Bill shape and size influence their feeding behavior and are also important features for identification. The brilliant iridescent colors displayed by many species, particularly among males, are the result of feather structure (vs. pigmentation) and are only visible from certain angles. Males of many species are quite vocal, though most produce little more than a monotonous series of thin, high-pitched notes. In a number of species, especially the hermits, groups of males (up to twenty or more at some sites) gather to sing for the purpose of attracting females. In these assemblages, known as leks, individual males are usually spaced apart by a meter or more (and can be frustratingly difficult to see amid the vegetation). Although hummingbirds are found in virtually every habitat in CR, the greatest species diversity occurs at middle elevations. World: 338, CR: 53

Hermits and sicklebill hummingbirds suspend their nest structures from the undersides of palms, heliconias, or other large-leaved plants. Most other hummers—like this Scintillant Hummingbird—build on top of a small branch or in the fork of a twig. The small cup has thick walls, relative to its overall size, in order to help insulate the eggs and nestlings while the mother is off foraging. Woven of spider threads and lined with soft materials, the nests are typically decorated on the outside with bits of moss and lichen that provide camouflage. In all species, two eggs are laid and the young fledge about six to seven weeks later. The female alone tends to the nesting chores.

White-necked Jacobin
Florisuga mellivora

male

female

5 in (13 cm). Even when male's posture hides its white nape band, the royal blue head and immaculate-white lower breast and remaining underparts are diagnostic. Females are easily confused, but show more spots on the breast than any similar species. Visits a variety of flowering trees and shrubs at all levels of forest edges and gardens; often seen zipping back and forth, catching insects in mid-air. The nest is built in the forest understory atop a horizontal palm frond, usually with another frond above for cover. Fairly common in wet lowlands and foothills; to 3,300 ft (1,000 m). Range: S Mexico to N Bolivia.

male

Green Hermit

Phaethornis guy

female

6 in (15 cm). This is the common large hermit of wet foothills and middle elevations, and the only local hermit that exhibits sexual dimorphism. The male is dark green-blue with a subtle facial pattern and white tips to the slightly longer central tail feathers, while the female has an obvious facial pattern, pale underparts, and elongated central tail feathers. Forages in lower levels of mature wet forest and forest edges. When visiting hummingbird feeders, it continues to hover while drinking and does not perch. Males lek in low, dense vegetation inside forest. Common in foothills and middle elevations; from 1,600 to 6,600 ft (500 to 2,000 m). Range: CR to E Peru.

Long-billed Hermit

Phaethornis longirostris

6 in (15 cm). The long, white-tipped, central tail feathers distinguish this large brownish hermit. Like other hermits, it visits heliconias, passionflowers, gingers, and other flowering plants in a manner known as "trap-lining," in which a route through the habitat is learned and repeated at intervals. Rarely comes to hummingbird feeders. Forages in lower levels of mature wet forest, advanced second growth, and forest edges. Males lek in low, dense vegetation inside forest, often near streams. Common in wet lowlands, and uncommon from 1,600 to 3,900 ft (500 to 1,200 m). Range: W Mexico to NW Peru.

Stripe-throated Hermit
Phaethornis striigularis

4 in (10 cm). The tapering, buff-tipped tail sets this small hermit apart from other brownish hummers. Visits a wide variety of plants, including those with rather small flowers such as the purple-flowering porterweed, so ubiquitously used as a hedge and attractant for hummingbirds. Rarely comes to hummingbird feeders. Forages low in mature forest, advanced second growth, forest edges and gardens. Males lek inside forest, using perches less than a meter above the ground. Common in wet lowlands and foothills, to 3,300 ft (1,000 m); uncommon from 3,300 to 5,200 ft (1,000 to 1,600 m) and in northern Pacific lowlands. Range: S Mexico to N Venezuela and W Ecuador.

Green Violetear
Colibri thalassinus

4 in (10 cm). Readily differentiated from other green hummingbirds by the purple ear coverts, which can be flared out in display and aggressive interactions. Also note the bluish tail with a darker subterminal band. Visits an assortment of flowers at all levels of forest edges and gardens; also a regular customer at hummingbird feeders. From high, exposed perches, males incessantly sing a typically two-note song that, once learned, is a good indicator of just how common this species is. Common in highlands, from Tilarán Cordillera south; from 4,600 ft (1,400 m) to timberline, though some descend to 3,000 ft (900 m) from March to Oct. Range: S Mexico to NW Argentina.

Purple-crowned Fairy
Heliothryx barroti

male

5 in (13 cm). A dazzling hummer with pure white underparts and outer tail feathers. Note how short the bill is on this relatively large hummingbird. Crown on male shows more purple than on female. Easily overlooked as it forages mostly at middle and upper levels of mature wet forest and tall second growth; most often seen at forest edges and in gardens, where it feeds at all levels, but does not visit hummingbird feeders. Uses its sharp bill to pierce the bases of large flowers, and thus extract their nectar. Fairly common from lowlands to middle elevations; to 5,200 ft (1,600 m). Range: SE Mexico to SW Ecuador.

Green-breasted Mango
Anthracothorax prevostii

male

female

5 in (13 cm). Male could be confused with the generally much more common Rufous-tailed Hummingbird (p. 100), but the tail is maroon and the breast has a dark wash (down its center) bordered by blue. Female is the only local hummer with a dark stripe down the center of underparts. Also note the entirely black bill. (In the Golfo Dulce-Osa region, this species is replaced by the very similar Veraguan Mango *A. veraguensis* [not illustrated].) Feeds at all levels in fairly open areas with some trees, forest edges, gardens, and mangroves. Fairly common in Palo Verde NP and Caño Negro regions, uncommon elsewhere in northern Pacific; also uncommon in Central Valley, Caribbean lowlands, and in central Pacific—south to about Uvita; rarely above 3,300 ft (1,000 m). Range: E Mexico to NW Peru.

Green Thorntail
Discosura conversii

male

female

Male 4 in (10 cm); female 3 in (8 cm). The white rump band is a field mark shared with the coquettes, so note the male's long tail feathers and the female's white moustache. (Be aware that several species of diurnal sphinx moths also have white rump bands, fly very much like hummingbirds, and even visit many of the same flowers!) In forest canopy visits trees with small flowers; forages down to lower levels at forest edges and in gardens, and comes to feeders. Very bee-like in flight. Fairly common in Caribbean foothills and middle elevations, from Tilarán Cordillera south, from 1,300 to 5,200 ft (400 to 1,600 m); rare visitor to adjacent lowlands. Range: CR to W Ecuador.

Black-crested Coquette
Lophornis helenae

male

female

3 in (8 cm). Fanciful male, with wiry crest feathers and frilled gorget, is quite distinctive. The female has pale underparts with some greenish dappling and lacks a distinguishing facial pattern. In forest canopy, visits trees with small flowers, forages down to lower levels at forest edges and in gardens, often coming to *Stachytarpheta*, but not known to visit feeders. Very bee-like in flight. Uncommon in northern Caribbean foothills and adjacent lowlands (south to watershed of Reventazón River), rare in Central Valley; to 3,900 ft (1,200 m). (The somewhat similar White-crested Coquette *L. adorabilis* [not illustrated] inhabits the Pacific lowlands and foothills.) Range: S Mexico to CR.

Green-crowned Brilliant
Heliodoxa jacula

male

5 in (13 cm). The small white spot behind the eye is a field mark shared with the similarly large Magnificent Hummingbird (p. 90); the two species occur together where their elevations overlap, so note the glittering blue spot on the male's throat, as well as overall green coloration, and the short white streak below the eye on the female, in addition to the rather speckled breast. Juvenile male has a rufous malar stripe. Visits all levels of wet montane forest, second growth, and forest edges; prefers to perch (vs. hover) when feeding; numerous at feeders. Common at middle elevations, from 2,300 to 7,200 ft (700 to 2,200 m); from Feb to June, occurs as low as 300 ft (100 m). Range: CR to W Ecuador.

juvenile male

female

Magnificent Hummingbird
Eugenes fulgens

male

female

5 in (13 cm). The purple crown and turquoise-blue throat set apart the male. The female is grayer below than any other large montane hummer and has a very long bill that droops slightly (the male's bill is straight). Frequents edges and openings in highland oak forest, also found in gardens; often a male will attempt to dominate a hummingbird feeder, warding off other comers. Occurs in Central and Talamanca Cordilleras, where it is common in highlands, from 5,900 ft (1,800 m) to timberline, and uncommon from 4,300 to 5,900 ft (1,300 to 1,800 m) on Caribbean slope. Range: SW US to W Panama.

throat colors displayed

Fiery-throated Hummingbird
Panterpe insignis

4 in (10 cm). If the spectacular array of colors on the forecrown, throat, and breast can be seen, identification is a cinch. However, these colors can be hard to discern given typical viewing conditions in the field, so note green back, violet-blue rump, and steely blue-black tail. Active at all levels of forests, forest edges, and clearings; visits feeders. Often very pugnacious; utters a screechy sputter. Common in Central and Talamanca Cordilleras, mostly above 6,600 ft (2,000 m); uncommon from Miravalles Volcano south through Tilarán Cordillera, above 4,900 ft (1,500 m). From March to July—throughout its range—occasionally descends to as low as 2,300 ft (700 m). Range: CR to W Panama.

Purple-throated Mountain-gem
Lampornis calolaemus

male

female

4 in (10 cm). The obvious white postocular stripe distinguishes both sexes as mountain-gems. Throat color is definitive on males, but females of this and the White-throated Mountain-gem (p. 92) are virtually identical. (The White-bellied Mountain-gem *L. hemileucus* [not illustrated] of Caribbean foothills and middle elevations is told apart by its white belly in both sexes.) Feeds at all levels of wet montane forest, second growth, forest edges, and gardens; numerous and aggressive at feeders. Common at middle elevations and in highlands, from Guanacaste Cordillera to northern end of Talamanca Cordillera; from 3,300 to 8,200 ft (1,000 to 2,500 m). From May to Sept, some descend to as low as 1,000 ft (300 m). Range: Nicaragua to central Panama.

White-throated Mountain-gem
Lampornis castaneoventris

male

4 in (10 cm). The male's white throat is diagnostic; also note the gray tail. (This species was formerly considered a Costa Rican endemic and was known as Gray-tailed Mountain-gem *L. cinereicauda*, but is currently lumped with the Panamanian form, which has a blue tail.) Female is virtually identical to the female Purple-throated Mountain-gem (p. 91). Forages at middle and lower levels of oak forests, forest edges, and gardens; regularly visits feeders. Fairly common in highlands of Talamanca Cordillera, from 5,900 ft (1,800 m) to timberline; from May to Sept, some descend to 4,900 ft (1,500 m). Range: CR to W Panama.

female

Magenta-throated Woodstar
Calliphlox bryantae

male

Male 4 in (10 cm); female 3 in (8 cm). The small size and overall coloration could cause confusion with the Volcano and Scintillant (p. 94) Hummingbirds, but the white patch on the side of the rump distinguishes both sexes. Also note male's long tail. Often feeds on low-growing flowers at forest edges, but tends to perch on high, exposed twigs; foraging flight is slow and insect-like, with body held horizontally and tail cocked up—which really sets it apart when at a feeder. Uncommon from 2,300 to 5,900 ft (700 to 1,800 m). Range: CR and W Panama.

female

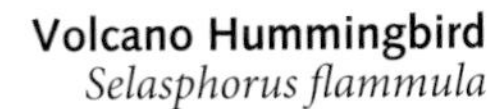

Volcano Hummingbird
Selasphorus flammula

male Poás race

male Irazú race

3 in (8 cm). This is the only tiny hummingbird above 2,400 m, but in the lower portions of its range it overlaps with the Scintillant Hummingbird (p. 94) and can be told apart by throat color in males (rosy in the northern half of the Central Cordillera (Poás), reddish-purple on Irazú and Turrialba Volcanoes, and purplish-green in the Talamancas) and the greenish central tail feathers of both sexes. Visits plants with small flowers at forest edges, gardens, and overgrown open areas; also occurs in paramo. Visits hummingbird feeders, but is often chased away by larger species. Occurs in Central and Talamanca Cordilleras; common above 5,900 ft (1,800 m); uncommon from 3,900 to 5,900 ft (1,200 to 1,800 m), mostly from March to July. Range: CR and W Panama.

male Talamanca race

female

Scintillant Hummingbird
Selasphorus scintilla

male

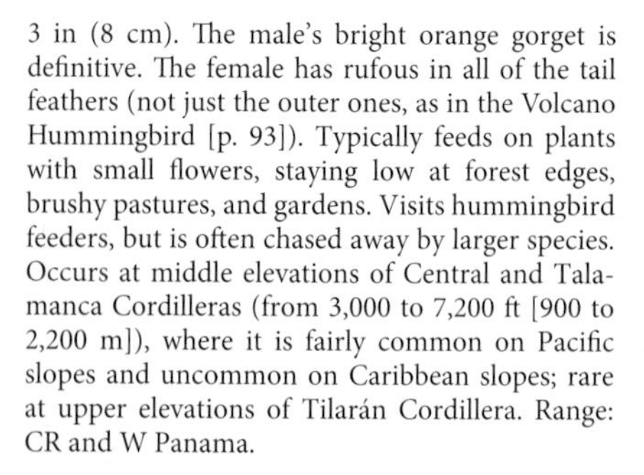

3 in (8 cm). The male's bright orange gorget is definitive. The female has rufous in all of the tail feathers (not just the outer ones, as in the Volcano Hummingbird [p. 93]). Typically feeds on plants with small flowers, staying low at forest edges, brushy pastures, and gardens. Visits hummingbird feeders, but is often chased away by larger species. Occurs at middle elevations of Central and Talamanca Cordilleras (from 3,000 to 7,200 ft [900 to 2,200 m]), where it is fairly common on Pacific slopes and uncommon on Caribbean slopes; rare at upper elevations of Tilarán Cordillera. Range: CR and W Panama.

female

Violet-headed Hummingbird
Klais guimeti

male

3 in (8 cm). The combination of an obvious white spot behind the eye and bluish-violet forecrown identifies both sexes of this small hummer. Feeds on a variety of plants with small flowers, in forest canopy and at lower levels of forest edges and gardens. When foraging, flies slowly, with body held horizontal and tail cocked up. Males form leks and sing from thin twigs fairly high up in forest gaps or at edge of tall second growth. Fairly common in Caribbean foothills and adjacent lowlands, to 3,300 ft (1,000 m); uncommon in southern Pacific, from 1,000 to 3,900 ft (300 to 1,200 m); uncommon around Golfo Dulce; casual in northern and central Pacific lowlands. Range: E Honduras to central Bolivia.

Violet Sabrewing
Campylopterus hemileucurus

male

6 in (15 cm). A definite crowd-pleaser, the large male is truly stunning and unmistakable. The duller female is similarly large with a decurved bill and much white in the outer tail feathers, and so should be fairly easy to ID. Found at lower levels of mature wet forest, often at openings and forest edges. Despite its size, tends not to be dominant at feeders. Males lek in low, dense vegetation inside forest. Common at middle elevations, from 3,300 to 7,900 ft (1,000 to 2,400 m); some descend to lower elevations from Nov to April, occasionally even to sea level. Range: S Mexico to W Panama.

Crowned Woodnymph
Thalurania colombica

male

female

4 in (10 cm). In dim light, the dazzling male can appear entirely dark, but the tail often shows a distinct fork. The female has a U-shaped line where the gray breast abuts the greenish belly. One of the most common forest hummingbirds in its habitat; the whirring of wings is often the first clue to its presence. Inhabits lower and middle levels of mature wet forest, second growth, and forest edges; often found at *Heliconia* flowers; visits feeders in or near forest. Common in wet lowlands and foothills, to 3,000 ft (900 m) on Caribbean slope and to 3,900 ft (1,200 m) in southern Pacific; rare north of Quepos on Pacific slope. Range: Belize to NW Venezuela.

Stripe-tailed Hummingbird
Eupherusa eximia

male

female

4 in (10 cm). The male is the only CR hummer to combine green underparts with a rufous wing patch (visible both in flight and perched) and white in the outer tail feathers. The female is similar, but has mostly whitish underparts. Inhabits mature wet montane forest canopy, but also feeds on flowering shrubs at forest openings, edges, and in gardens, where it seems especially fond of *Canna* flowers. The only local hummer that incorporates red lichen on the outside of its nest. Uncommon at middle elevations, from 2,600 to 7,200 ft (800 to 2,200 m); note some seasonal altitudinal movements (up to 1,600 ft [500 m] higher or lower than its normal range). Range: S Mexico to W Panama.

Black-bellied Hummingbird
Eupherusa nigriventris

male

3 in (8 cm). The male's black face and underparts are unique among local hummers. Female resembles female Stripe-tailed Hummingbird, but has reduced rufous wing patch and entirely white outer three tail feathers. Prefers mature wet montane forest and forest openings and edge. Uncommon at middle elevations on Caribbean slope from Central Cordillera south, from 3,000 to 6,600 ft (900 to 2,000 m); moves lower from April to Sept (down to 2,000 ft [600 m]). Range: CR and W Panama.

female

Coppery-headed Emerald

Elvira cupreiceps

male

female

3 in (8 cm). One of just three species found only in mainland CR. The male's coppery crown and rump are diagnostic. The female resembles several other small female hummers that are green above, pale below, and flash white in the outer tail feathers, but note the decurved bill. Visits trees and shrubs with small flowers at all levels in mature wet montane forest and forest edges; also comes to hummingbird feeders. Males form small leks at middle levels of forest edges. Fairly common at middle elevations on Caribbean slope, north of Reventazón River, from 2,000 to 4,900 ft (600 to 1,500 m); also fairly common on Pacific slope of Guanacaste and Tilarán Cordilleras, from 3,900 to 4,900 ft (1,200 to 1,500 m). Range: Costa Rica.

Snowcap
Microchera albocoronata

male

3 in (8 cm). One of CR's smallest hummingbirds. The male's wine-red coloration and gleaming white crown render it unmistakable. The female resembles several other small female hummers, but note the white in the outer tail feathers, lack of green spotting on the plain underparts, and the short, straight bill. Forages on small flowers in canopy and edges of both mature wet forest and second growth. Uncommon on Caribbean slope, from 1,000 to 3,000 ft (300 to 900 m); some descend to adjacent lowlands during latter half of year. Range: S Honduras to central Panama.

female

juvenile male molting into adult plumage

male

Mangrove Hummingbird

Amazilia boucardi

4 in (10 cm). One of just three species found only in mainland CR—and, then, only in or near Pacific coast mangroves. In its restricted habitat it is the only hummer likely with white lower underparts; also note green dappling on sides. The male has bluish speckling on throat and sides of neck. Forages at lower and middle levels of mangrove swamps and adjacent vegetation; prefers nectar of flowers from the buttressed Tea Mangrove *Pelliciera rhizophorae*. Uncommon in mangroves, from Gulf of Nicoya south to Golfo Dulce, and also around Tamarindo. Range: Costa Rica.

Steely-vented Hummingbird

Amazilia saucerrottei

4 in (10 cm). The coppery rump is key to distinguishing this from other mostly green hummingbirds. The dark bluish tail can sometimes look very bright. Feeds at all levels of forest openings and edges, second growth, and gardens. Common in northern Pacific; to 3,900 ft (1,200 m). Fairly common in northern central portion of Caribbean slope, Central Valley (east to Turrialba), and Dota region (south to Fila Costera, east of Dominical); to 5,200 ft (1,600 m). Range: W Nicaragua to W Venezuela.

Rufous-tailed Hummingbird
Amazilia tzacatl

4 in (10 cm). The combination of a green throat and breast with a rufous rump and tail is diagnostic. This is the common hummer found in most human-altered habitats (rarely inside forested areas) and a good species to become familiar with in order to compare it with other species, since it is medium-sized with a medium-length bill. Typically very aggressive, it often engages in chases and is especially bullying toward smaller species. Visits many kinds of flowers and readily comes to feeders. Widespread; common in wet habitats, to 5,200 ft (1,600 m); but uncommon from 5,200 to 7,200 ft (1,600 to 2,200 m). Uncommon in northern Pacific lowlands. Range: E Mexico to W Ecuador.

Cinnamon Hummingbird
Amazilia rutila

4 in (10 cm). Quite similar to the Rufous-tailed Hummingbird in both appearance and habits (and more common in most of its CR range) but has entirely pale-rufous underparts; female Purple-throated Mountain-gem (p. 91) has white postocular stripe and lacks rufous in tail. Frequents forest edges, second growth, and gardens; comes to feeders. Fairly common in northern Pacific and western Central Valley; to 3,300 ft (1,000 m). Casual on Caribbean slope, where reported from Puerto Viejo de Sarapiquí and Turrialba. Range: W Mexico to NW CR.

Pigeons, Doves (COLUMBIDAE). The members of this cosmopolitan family should be recognizable as such to most people, thanks, in large part, to the ubiquitous Rock Pigeon, which was domesticated even before the chicken. The full-breasted appearance is due to powerful flight muscles—a fact that results in many species being shot for both "sport" and meat. Slightly more than half of the species are predominantly seed eaters, the rest consume mostly fruit. Most use twigs to construct a shallow nest; placement may be on the ground or on a cliff, or above the ground in vegetation. Typically one or two eggs are laid and the young are tended by both parents, at first being fed only nutritious "crop milk"—a unique substance produced from cells in the crop (a pouch near the throat). World: 329, CR: 25

male

Pale-vented Pigeon

Patagioenas cayennensis

12 in (30 cm). If the pale underparts are not visible, note the red iris and black bill. Stays in upper levels of trees in nonforest habitats or at forest edges; regularly perches on utility wires. Feeds on both seeds and small fruit. In display flight, glides with wings held up in a V. Coos with a three-part *hu-hu-huUUu*, sometimes introduced by a low *wuuu*. Common in wet lowlands; to 2,000 ft (600 m). Range: SE Mexico to N Argentina.

female

Red-billed Pigeon

Patagioenas flavirostris

12 in (30 cm). A misleading name, as only the cere and the basal third of the bill are pink-red; the distal portion of the bill is creamy-white. Note the dull ruddy color of the head, underparts, and shoulders, and the otherwise gray wings and back. Individuals or small groups are typically seen perched in the tops of trees or flying swiftly past. Forages for small fruits and flower buds in middle and upper levels of trees in nonforest habitats. Coos with a forlorn *huUUu-hu-hu-huuu*. Fairly common in northern Pacific, Central Valley, and middle elevations of Caribbean slope (to 6,600 ft [2,000 m]); uncommon elsewhere within mapped range. Range: S Texas to central CR.

Band-tailed Pigeon

Patagioenas fasciata

14 in (36 cm). The most commonly seen highland pigeon. The yellow bill and white band on nape are diagnostic. Typically perches in canopy at edges of oak forests and gardens; also seen flying swiftly past; often in flocks. Consumes acorns and berries. Makes a rather guttural *huhuUU, hu-huUU*; also produces loud wing-flapping sound. Common in highlands from timberline down to about 4,600 ft (1,400 m); occasionally to 1,600 ft (500 m), or lower. Range: SW Canada to N Argentina.

Short-billed Pigeon
Patagioenas nigrirostris

10 in (25 cm). The lack of field marks is the best field mark on this arboreal pigeon. However, very similar to the highland Ruddy Pigeon *P. subvinacea* (not illustrated) in appearance and habits (and they occasionally overlap at lower middle elevations); best told apart by differently accented songs: *hu, HU, hu-HU* in Short-billed and *hu, hu-HU, huu* in Ruddy. More often heard than seen as it tends to stay in forest canopy. Feeds on fruits, especially mistletoe and *Cecropia*; does not flock. Common in wet forest habitats, to 3,900 ft (1,200 m); uncommon in Caño Negro region. Range: SE Mexico to NW Colombia.

White-winged Dove
Zenaida asiatica

11 in (28 cm). The white wing stripe is very obvious in flight and also visible on perched birds. Found in fairly open, nonforested areas; often perches on wires and feeds along roadsides, picking up fallen seeds and spilled grain. Quite commensal with humans. Varied calls include a throaty "who-cooks-for-you." Common resident in northern Pacific and across Central Valley (east to Paraíso); to 4,600 ft (1,400 m). Still rare on northern Caribbean slope, but established in La Fortuna. Resident population joined by migrants from Nov to May. Range: SW US to W Panama.

Inca Dove
Columbina inca

8 in (20 cm). The combination of scaled pattern, black bill, and long tail with white outer feathers sets it apart from other small ground-doves. Individuals or pairs are typically found foraging for seeds on the ground in fairly open areas. From an above-ground perch, gives an incessant *huuu-huup*. Common in northern Pacific and western Central Valley; still extending its range south along Pacific coast and east of Cartago; to 4,600 ft (1,400 m). Range: SW US to SW CR.

Ruddy Ground-Dove
Columbina talpacoti

male

7 in (18 cm). In humid regions, easily the most common of the small, mostly terrestrial doves. The male's coloration facilitates ID, but the browner female resembles several other dove species. A distinctive characteristic is the rufous rump, visible in flight. Pairs or small groups (often with other small doves) forage for seeds on ground in open areas. Repeats a low *hu-hUUp*. Common in wet lowlands and in eastern Central Valley, uncommon to rare in northern Pacific; to 4,600 ft (1,400 m). Range: S Mexico to central Argentina.

female

White-tipped Dove
Leptotila verreauxi

11 in (28 cm). A rather plainly attired, pigeon-sized dove; the light-blue orbital skin sets it apart from other similar members of the genus. Pairs or individuals forage for fallen fruit and seeds on the ground in light woodlands, gardens, and roadsides; generally avoids mature forest habitats. Utters a low, mournful *whuuu*, lasting about one second (at intervals of six to seven seconds). Common on Pacific slope and in Central Valley, to 7,200 ft (2,200 m); uncommon (though increasing) in Caribbean lowlands. Range: S Texas to central Argentina.

Pacific race

Gray-chested Dove
Leptotila cassini

11 in (28 cm). Very similar to other *Leptotila* doves, but differentiated by the golden-brown nape (less obvious on Caribbean race). Also note the red orbital skin. Pairs or individuals forage for fallen fruit and seeds on the ground in wet forests, at forest edges, and in shaded gardens, walking with the head-bobbing motion that is characteristic of most members of the family. Its forlorn call, similar in quality to that of the White-tipped Dove, rises then falls slightly, lasting about 1.5 seconds. Fairly common in wet lowlands and foothills; to 3,900 ft (1,200 m). Range: Guatemala to N Colombia.

Buff-fronted Quail-Dove
Zentrygon costaricensis

11 in (28 cm). Quail-doves are handsome, essentially terrestrial, members of the pigeon family. Unlike their namesakes, the quails, they do not form coveys. This species—the most common quail-dove at higher elevations—is aptly named for its distinguishing field mark. Generally encountered individually; forages for fallen fruit and seeds on the ground in wet montane forests; commonly walks on trails. Incessantly repeats a rising *wha, wha, wha, wha...* . Uncommon at middle elevations and in highlands; from 3,900 to 9,800 ft (1,200 to 3,000 m). Range: CR and W Panama.

Olive-backed Quail-Dove
Leptotrygon veraguensis

9 in (23 cm). The darkest quail-dove in CR, it can be distinguished by the white cheek stripe and broad, dark-gray malar stripe. Pairs or individuals forage on ground in wet forest; commonly walks on trails. Gives a hoarse, somewhat amphibian-like *whu*. Uncommon in wet Caribbean lowlands; to 1,600 ft (500 m). Range: CR to NW Ecuador.

Parrots (PSITTACIDAE). Distributed principally throughout the African and New World tropics and subtropics, members of this family are readily distinguished by their characteristic bill shape. Identification to species is aided by coloration on the head, though getting a good view of parrots in the wild can be notoriously difficult as they are often either silhouettes in flight or camouflaged amid green vegetation. Listen for their vocalizations or the sound of falling pieces of fruit to help locate them. Ripening seeds are the primary dietary item and the powerful bill is designed for ripping through hard seed coats. Pairs apparently mate for life and nests are usually in tree cavities, though some smaller species also excavate in termitaries. Sadly, persecution for the pet trade, combined with habitat loss, threatens numerous species. World: 167, CR: 17

Orange-chinned Parakeet

Brotogeris jugularis

7 in (17 cm). The most numerous and widespread member of the family in CR, this small parakeet can be recognized by its short, pointed tail and brownish shoulders (since the small orange spot beneath the bill is usually difficult to see). Pairs and noisy flocks prefer trees in clearings and at forest edges. Produces a near constant, shrill, harsh chatter. Common and widespread in lowlands and foothills; to 3,900 ft (1,200 m). Has expanded its local distribution due to continued deforestation. Range: SW Mexico to W Venezuela.

Crimson-fronted Parakeet
Psittacara finschi

11 in (28 cm). The most common parakeet in the Central Valley, it travels in flocks of a dozen up to 100 or more individuals. Together, the red forecrown and long pointed tail are diagnostic. Juvenile has mostly green forecrown. Frequents gardens, shaded coffee plantations, and forest edges; often found in *Erythrina* and palm trees. Produces a boisterous chattering *klee-klee-chee-chee...* (somewhat similar to calls of other parakeets). Common on the Caribbean and S Pacific slopes, and across the Central Valley; to 5,900 ft (1,800 m). Range: S Nicaragua to S Panama.

Orange-fronted Parakeet
Eupsittula canicularis

9 in (23 cm). The only long-tailed parakeet in the dry forest; the orange forecrown and prominent eye ring confirm the ID. Flocks visit forests, agricultural areas, and mangroves. Usually nests in termitaries. Screeching call similar to that of Crimson-fronted Parakeet. Fairly common in northern Pacific, uncommon in western Central Valley; to 3,300 ft (1,000 m). Range: W Mexico to NW CR.

Olive-throated Parakeet
Eupsittula nana

9 in (23 cm). Has entirely green head and dull olive throat. Small flocks feed in fruiting trees in gardens and at forest edge. Calls are higher pitched and not as harsh as those of Crimson-fronted Parakeet. Fairly common throughout Caribbean lowlands; rarely up to 2,300 ft (700 m). Range: NE Mexico to W Panama.

Great Green Macaw
Ara ambiguus

33 in (84 cm). A beautiful study in colors; this large bird is unmistakable. Pairs or small flocks fly low over the forest canopy, often revealing their presence with their harsh calls (loud *aak, raak* can be heard at a great distance); seasonal movements are related to ripening of *Dipteryx* fruits, locally known as *almendro*. Increasingly uncommon in northeastern Caribbean lowlands, rare south of Limón; occasionally to 2,600 ft (800 m). Range: E Honduras to W Ecuador.

Scarlet Macaw
Ara macao

35 in (90 cm). This unmistakable and spectacular bird forages in the forest canopy and at edges, and is regularly found feeding in West Indian Almond (*Terminalia cattapa*) trees that grow along beaches; typically in pairs or small groups composed of various pairs. Roosts in mangroves. Flying birds announce themselves with a raucous, harsh *raaak, raaak*. This species is now readily found only in the Carara-Tarcoles area and on the Osa Peninsula. A few birds still remain in the Palo Verde area, and recent sightings in the NE Caribbean lowlands suggest it may be recolonizing areas where it has not been seen for at least fifty years. Range: SE Mexico to Bolivia.

Brown-hooded Parrot
Pyrilia haematotis

9 in (23 cm). Brownish head with white orbital ring and red ear patch is distinctive. In flight, note the red axillars. Forages in forest canopy and at forest edge; flies quickly, with rocking motion and strong wingbeats. Call is a high-pitched *chreea, cheea*. Fairly common in wet lowlands and at middle elevations; to 4,900 ft (1,500 m). Range: S Mexico to NW Colombia.

Note red axillars on flying birds.

White-crowned Parrot
Pionus senilis

10 in (25 cm). The large white forecrown patch (extending to the top of the otherwise dull slate-blue head) distinguishes it from all other local species. Flies with very deep wingstrokes. Small flocks frequent gardens and forest edges; commonly perches on the tips of unopened, vertical palm fronds. Very noisy, especially in flight, making a shrill chatter. Common in wet lowlands and middle elevations, uncommon in Central Valley; to 5,900 ft (1,800 m). Range: SE Mexico to W Panama.

[The following four species belong to the genus *Amazona*, considered the "classic" parrots (i.e., the kind pirates have on their shoulders). Body plumage is very similar among species, so markings on the head are crucial for ID; although juveniles generally lack these, there are usually adults nearby to help sort things out. They have broad wings, square tails, and fly with such stiff, shallow wingbeats that it almost appears they shouldn't stay airborne. Where populations are healthy, they gather in treetops in the late afternoon prior to roosting and loudly call back and forth.]

White-fronted Parrot
Amazona albifrons

10 in (25 cm). This is the smallest of the local *Amazona* and easily told apart by the white forecrown, sky-blue crown, and red around the eye. It is also the only one that shows red on the leading edge of the wing in flight. Found in both wooded and open areas. Utters a raucous *ak-ak-ak-ak-ak*. Common in northern Pacific; rare, though perhaps increasingly less so, in western Central Valley and northern Caribbean foothills; to 3,900 ft (1,200 m). Range: NW Mexico to NW CR.

Red-lored Parrot
Amazona autumnalis

13 in (33 cm). Virtually identical to the Mealy Parrot except for the red forehead. (Note that some individuals also have yellowish cheeks.) Prefers forest edges and gardens; more tolerant of deforestation than Mealy Parrot. On average, its varied calls are higher-pitched than those of Mealy Parrot; listen for its metallic *klink, klink, klink*, especially when it is in flight. Common in wet lowlands and foothills; to 3,300 ft (1,000 m). Range: E Mexico to NW Brazil.

Yellow-naped Parrot
Amazona auropalliata

14 in (36 cm). Similar to the Mealy Parrot, but has a bright yellow nape (absent on juvenile); there is almost no overlap in range, except in the Tarcoles-Carara area. Forages high in trees. Makes a wide variety of deep utterances, without all the squawking of others in the genus. Unfortunately, its value as a pet bird has led to population decline. Uncommon in northern Pacific; to 2,000 ft (600 m); birds seen in Central Valley may be escaped cage birds. Range: S Mexico to NW CR.

Mealy Parrot
Amazona farinosa

15 in (38 cm). Virtually identical to the Red-lored Parrot, but has blackish ceres and lacks red forehead. More dependent on forested tracts than Red-lored Parrot, although it also frequents forest edges and gardens, where the two can occur together at good fruiting trees. Among its many vocalizations, notice a deep *cheyup, cheyup*. Common in wet lowlands; to 2,000 ft (600 m). Range: SE Mexico to E Brazil.

Cuckoos (CUCULIDAE). The term *cuckoo* derives from the song of the Old World Common Cuckoo *Cuculus canorus*, made famous by its use in clocks. While none of the New World species makes a similar sound, their vocalizations are important for finding these often concealed birds. Most species are insectivorous, many specializing on caterpillars—even hairy or spiny varieties containing noxious chemicals! The larger ground-dwelling species also consume small vertebrates. The majority of the family exhibits standard monogamous pair nesting; however, two extreme variants have evolved in some species: brood parasitism and communal nesting. World: 145, CR: 12

Squirrel Cuckoo
Piaya cayana

18 in (46 cm). Its long tail, overall size, and manner of hopping along branches truly impart a squirrel-like impression. Not likely confused with any other bird. Found in gardens, second growth, and at forest edges; prefers caterpillars, but also eats other insects and small lizards. Flies from one tree to the next with a few wing flaps and a long glide. Among its various calls is an arresting *IK-weyeew*. In breeding season (from Jan to Aug), gives long series of dry *whip* notes. Common and widespread; to 7,500 ft (2,300 m). Range: N Mexico to N Argentina.

Striped Cuckoo
Tapera naevia

12 in (30 cm). Somewhat resembles a large, slender sparrow, but note decurved upper mandible ending in a hooked tip (vs. a conical bill). Inhabits brushy fields and second growth; forages in undergrowth and on the ground for large insects. Usually sings from an exposed perch, whistling two clear notes, the second slightly higher, all the while raising and lowering its crest feathers. One of just three New World cuckoos that is a brood parasite; the female lays her eggs in other species' nests, especially domed nests of spinetails, wrens, and sparrows. Fairly common in Caribbean lowlands, southern Pacific, and western Central Valley; to 3,600 ft (1,100 m). Uncommon around Gulf of Nicoya and on Nicoya Peninsula. Range: SE Mexico to N Argentina.

Lesser Ground-Cuckoo
Morococcyx erythropygus

0 in (25 cm). If seen, the color and pattern around the eye is diagnostic. Stealthily keeps to thickets nd brushy fields, foraging on the ground for insects, and capable of running swiftly, like a miniaure roadrunner. It occasionally perches in the open on a rock, fence post, or low branch, especially fter a rain. At intervals, delivers a loud, long series of trilled whistles that accelerates then gradually lows. Fairly common in northern Pacific and in western Central Valley; to 3,900 ft (1,200 m). Range: V Mexico to NW CR.

Rufous-vented Ground-Cuckoo
Neomorphus geoffroyi

19 in (48 cm). This "prize" species is virtually unmistakable. Dwells in understory of wet forest. Follows army-ant raids and, reportedly, peccaries to feed on the insects that they flush out. Rare in Caribbean lowlands and foothills, also on Pacific slope of Guanacaste Cordillera; to 3,000 ft (900 m). Range: Nicaragua to SE Brazil.

Groove-billed Ani
Crotophaga sulcirostris

12 in (30 cm). Best told apart from the somewhat larger Smooth-billed Ani *C. ani* (not illustrated) of the S Pacific, as well as all other essentially black birds, by the striations on the bill. Small groups forage on the ground and in low vegetation in fields and gardens; often follows cattle to eat the insects they stir up. Common call is a tinkling *TEE-ho, TEE-ho*. The anis are comunal nesters with up to four pairs laying a total of as many as a dozen or more eggs in a shared nest. Th dominant male does the largest share of incubation, while his mate does the least. Common an widespread; to 4,900 ft (1,500 m); rare in southern Pacific (where essentially replaced by the Smootl billed Ani). Range: S Texas to N Argentina.

Trogons (TROGONIDAE). Trogons and quetzals form a pantropical family of essentially forest species noted for their bright, contrasting color patterns (more dramatic in males). In many species, however, the true key to identification lies in bill, eye, and/or eye ring color, along with the undertail pattern. Of course, vocalizations are also important and are often how one first realizes a trogon is nearby, since they characteristically perch quite still in an upright position. They make quick sallies to pluck fruit or grab insects or small vertebrates, briefly hovering as they do so. All are cavity nesters that excavate their own nests. Sites include rotting trunks, arboreal termite nests, and even paper wasp nests! Both members of the pair participate in all the nesting duties. World: 44, CR: 10

Resplendent Quetzal
Pharomachrus mocinno

male

female

14 in (36 cm). The bird everyone wants to see—and when you do, you'll know why! The male is unmistakable. The four greatly elongated "tail" feathers are actually uppertail coverts that extend up to 30 in (76 cm) beyond the true tail tip. After breeding, these feathers are molted, and from Aug to Nov it is common to find male quetzals without them. The female is the only CR trogon with a gray lower breast. Inhabits middle levels of mature wet montane forest, forest edges, advanced second growth, and even enters gardens when ripe fruit is available; feeds mostly on members of the avocado family. The call is a throaty *kyow, kyow*; in flight, gives a brisk *wicka-wicka*. Fairly common in highlands of Tilarán, Central, and Talamanca Cordilleras (on both Pacific and Caribbean slopes); from 4,600 to 9,800 ft (1,400 to 3,000 m). In the northern part of their CR range (Tilarán Cordillera), they are altitudinal migrants; most of the year they occur from 4,600 to 5,200 ft (1,400 to 1,600 m), but from about Sept to Nov, they descend the Caribbean slope, where most then occur from 1,600 to 2,600 ft (500 to 800 m). Range: S Mexico to W Panama.

Slaty-tailed Trogon
Trogon massena

male

female

12 in (30 cm). The all-gray undertail is diagnostic, as all other CR trogons show some white. Note that females on the Pacific slope have all-dark bills and could be confused with female Baird's Trogons (p. 120), but the unpatterned undertail and wing coverts are definitive. Forages in middle and upper levels of mature wet forest, though often lower at forest edges and in gardens; sometimes accompanies mixed species flocks. Typical call is a series of monotonous barking notes (*aah, aah, aah, aah…*), lower pitched than that of Gartered Trogon (p. 120). Common in wet lowlands and foothills to 3,900 ft (1,200 m). Range: SE Mexico to NW Ecuador.

Collared Trogon
Trogon collaris

male

female

10 in (25 cm). Virtually identical to the Orange-bellied Trogon, apart from the belly color, which is red (slightly duller in female). (The somewhat similar Elegant Trogon *T. elegans* [not illustrated] is an uncommon inhabitant of dry forests in the northern Pacific lowlands and foothills.) Found in middle levels of mature wet montane forest, forest edge, and gardens. The call is a series of sweet, down-slurred *caow, caow* notes, with several seconds of silence between each couplet. Fairly common in central and southern highlands, from 2,600 to 9,200 ft (800 to 2,800 m); descends to foothills on both slopes, from June to Dec. Range: Central Mexico to N Bolivia.

Orange-bellied Trogon
Trogon aurantiiventris

male

female

10 in (25 cm). Respective sexes virtually identical to the Collared Trogon in every aspect, including voice, habitat, and habits. The only difference is the color of the belly, which is orange (pale orange in female). Common in wet northwestern highlands (where Collared Trogon is absent), uncommon farther south, where the two species' ranges overlap; from 2,300 to 7,500 ft (700 to 2,300 m). Range: CR and western Panama.

Black-headed Trogon
Trogon melanocephalus

male

female

11 in (28 cm). This is the most commonly encountered trogon in the dry NW region of the country, where the fairly similar Gartered Trogon (p. 120) also occurs. The pale blue eye ring and large white tips to the outer tail feathers set apart the male. On the female, note the unbarred wing coverts. Frequents lower and middle levels of dry and gallery forests, forest edges, and gardens. The call is an accelerating series of chucks, rising in pitch and then terminating with several quickly descending notes. Common in northern Pacific lowlands and foothills and in Caño Negro region, rare in western Central Valley; to 2,600 ft (800 m). Range: E Mexico to CR.

Baird's Trogon
Trogon bairdii

male

female

11 in (28 cm). The male's orange belly and entirely white undertail are a unique combination. The female resembles the female Slaty-tailed Trogon (p. 118), but the dark gray undertail and wing coverts are finely barred with white. Forages from lower to upper levels of mature wet forest. The chuckling call begins at a moderate pace, then rises in both pitch and tempo before slowing and dropping. Fairly common in lowlands and foothills of southern Pacific; to 3,900 ft (1,200 m). Range: CR and W Panama.

Gartered Trogon
Trogon caligatus

male

female

9 in (23 cm). Formerly known as Violaceous Trogon, this is the most widespread trogon in CR. The male is readily distinguished by the yellow eye ring. The female has a whitish, eliptical eye ring and fine white barring on the wing coverts (compare with female Black-headed Trogon [p. 119]). Not a true forest species; it occurs in gallery forest, at forest edges, and gardens, usually at middle or upper levels; even perches on utility wires. The call—often given from a high perch—is a series of even-pitched notes (*kyew, kyew, kyew...*) that are higher pitched than those of the Slaty-tailed Trogon (p. 118). Common in wet lowlands and foothills, uncommon in drier areas; to 4,600 ft (1,400 m). Range: S Mexico to NW Peru and W Venezuela.

Black-throated Trogon
Trogon rufus

9 in (23 cm). The combination of a pale blue eye ring, green upperparts, and a yellow belly identifies the male. The female is the only CR trogon that has brown upperparts and a yellow belly. A true forest species, it favors lower and middle levels of mature wet forest and adjacent advanced second growth. The call—more whistled than that of other trogons—is a slow, deliberate series of two to five notes, reminiscent of a Chestnut-backed Antbird (p. 149), but all notes even-pitched. Common in wet lowlands and foothills; to 3,900 ft (1,200 m). Range: SE Honduras to NE Argentina.

male

female, dorsal view

female, ventral view

Motmots (MOMOTIDAE). It is assumed that this neotropical family evolved in what is now Mexico and Central America, as this is where the greatest species diversity lies. Handsome birds of forested habitats, most species have racquet-tipped tails (although when freshly molted, the tails are fully feathered) that they often swing from side to side as they perch (otherwise motionlessly) while scanning for prey. Large insects, small vertebrates, and occasional fruit form the diet. Their vocalizations are a prominent feature in the avian chorus at daybreak. Pairs excavate a horizontal burrow up to a meter or more deep in a riverbank or roadcut. World: 13, CR: 6

Blue-crowned Motmot

Momotus coeruliceps

ventral view

dorsal view

16 in (41 cm). The electric-blue ring around the crown differentiates this species, which can occur inside mature forest but seems to prefer forest edges and shady gardens, where it usually perches on fairly low branches. Often takes dust baths near dusk. Typically calls with a slow series of low, doubled notes (*hoop-hoop, hoop-hoop*...), which are the origin of the term *motmot*. Common in the Central Valley, southern Pacific slope, and middle elevations, to 6,600 ft (2,000 m); uncommon in evergreen patches of northern Pacific lowlands. Range: NE Mexico to W Panama.

Broad-billed Motmot
Electron platyrhynchum

12 in (30 cm). The blue-green chin is the field mark that distinguishes it from the much larger, but sympatric, Rufous Motmot. In the Arenal area, there have been several cases of this species breeding with the closely related Keel-billed Motmot. Inhabits lower and middle levels of mature wet forest. The call is a throaty *awnk*. Common in Caribbean lowlands and foothills; to 3,300 ft (1,000 m). Range: E Honduras to N Bolivia.

Keel-billed Motmot
Electron carinatum

13 in (33 cm). A mostly olive-green motmot with a rufous forehead and short, blue superciliary. Rare resident of Caribbean foothills along the Guanacaste and Tilarán Cordilleras; from 1,000 to 3,000 ft (300 to 900 m). Also known from the Sarapiquí region. Inhabits lower levels of mature wet forest and advanced second growth. Voice identical to that of Broad-billed Motmot, with which it interbreeds in CR. Range: SE Mexico to CR.

Rufous Motmot

Baryphthengus martii

18 in (46 cm). The large size, very long tail, and rufous on undersides extending all the way to the belly distinguish it from the similar Broad-billed Motmot (p. 123). Inhabits lower and middle levels of mature wet forest. Song is a deep, bubbling, rolling hooting that is most often heard at dawn. Fairly common in Caribbean lowlands and foothills; to 3,300 ft (1,000 m). Range: Honduras to Bolivia.

Turquoise-browed Motmot

Eumomota superciliosa

13 in (33 cm). Rust, olive, sky blue, and a bit of black have joined to produce a truly stunning bird. No other motmot has such a lengthy exposed shaft above the racquet tips. Inhabits dry and humid forests, forest edges, and even tree-lined roadsides (where it regularly perches on utility wires). The call is a hoarse *qwahk*. The common motmot of northern Pacific lowlands and foothills, rare in western Central Valley; to 3,000 ft (900 m). Range: S Mexico to Costa Rica.

Kingfishers (ALCEDINIDAE). Most species of this cosmopolitan family reside in the Old World, where plumages and behavior are more diverse. All six New World species can be found in CR, though the Belted Kingfisher *Megaceryle alcyon* (not illustrated) is a NA migrant. Species in CR feed primarily on fish, obtained by headlong plunges into water, with occasional other aquatic organisms taken in the same manner. Prey items are often beaten on a branch prior to being swallowed headfirst. Pairs excavate burrows in riverbanks, with nesting coinciding with periods of low water levels. Strangely, the adults do not remove waste from the nest as the young are growing. World: 95, CR: 6

Ringed Kingfisher

Megaceryle torquata

16 in (41 cm). The largest of the New World kingfishers; it is easily differentiated by the blue-gray upperparts and mostly rufous underparts. Found along rivers and streams, as well as lakes, lagoons, and large ponds, where it can typically be seen on relatively high perches, including wires. Tends to fly higher than other kingfishers, often giving a loud, rasping *krek* in flight. Also utters a loud, even rattle. Widespread and fairly common in lowlands and Central Valley; to 4,900 ft (1,500 m). Range: S Texas to Tierra del Fuego.

female

male

Amazon Kingfisher
Chloroceryle amazona

male

female

11 in (28 cm). The male is like a large version of the Green Kingfisher, but note the absence of white spotting on the wings; in flight, shows no white in the tail. The female can be recognized by the green markings on the sides of the breast (vs. a complete breast band). Inhabits larger rivers and streams, ponds, lakes, and lagoons. Sometimes hovers above water briefly before diving. Calls with a single harsh *chirt*, which is sometimes repeated in rapid succession to make a rattle. Fairly common in lowlands, and up to 3,900 ft (1,200 m) in Reventazón River watershed and in southern Pacific. Range: S Mexico to Argentina.

Green Kingfisher
Chloroceryle americana

male

female

7 in (18 cm). Both sexes similar to the larger Amazon Kingfisher, but have obvious white spotting on the wings and, in flight, show white on the outer tail feathers. The female has two green breast bands. Typically uses low perches, even rocks in waterways, and flies low along watercourses. Potentially encountered at any suitable wetland site (e.g., rainwater pools, small streams, and wide rivers). The ticking call resembles the sound made by striking two pebbles against each other. Common in lowlands and foothills; to 3,900 ft (1,200 m). Range: SW US to N Chile and N Argentina.

American Pygmy Kingfisher

Chloroceryle aenea

male

female

5 in (13 cm). By far the smallest CR kingfisher, perhaps more resembling an overstuffed hummingbird. The white underbelly and vent on otherwise pale rufous underparts is a unique combination. Favors swamp forests, mangroves, and shaded streams; occasionally found in more exposed situations. Generally perches within a meter of the water surface; also sometimes catches passing flying insects on the wing. The call is like a soft rendition of the Green Kingfisher's. Uncommon in lowlands; to 2,000 ft (600 m). Range: S Mexico to N Bolivia.

Puffbirds (BUCCONIDAE). A strictly neotropical family, the puffbirds show their greatest diversity in the Amazon basin. Characterized by large heads, stout bills, and short legs. They usually appear rather squat as they sit immobile, with an occasional turn of the head. This typical sit-and-wait predator strategy belies the swiftness with which they fly when attacking prey (large arthropods and small vertebrates). On returning from a successful foray, the bird usually beats the prey item against the branch prior to swallowing it. Pairs excavate nests either in arboreal termite nests or in the ground. World: 36, CR: 5

White-whiskered Puffbird

Malacoptila panamensis

male

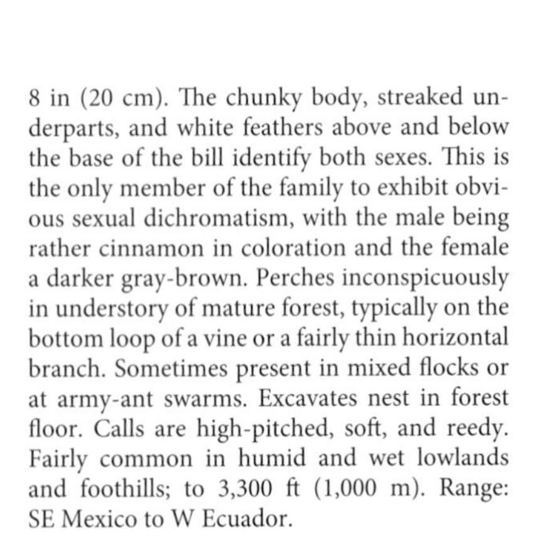

8 in (20 cm). The chunky body, streaked underparts, and white feathers above and below the base of the bill identify both sexes. This is the only member of the family to exhibit obvious sexual dichromatism, with the male being rather cinnamon in coloration and the female a darker gray-brown. Perches inconspicuously in understory of mature forest, typically on the bottom loop of a vine or a fairly thin horizontal branch. Sometimes present in mixed flocks or at army-ant swarms. Excavates nest in forest floor. Calls are high-pitched, soft, and reedy. Fairly common in humid and wet lowlands and foothills; to 3,300 ft (1,000 m). Range: SE Mexico to W Ecuador.

female

White-necked Puffbird
Notharchus hyperrhynchus

10 in (25 cm). The large head and bill suggest a kingfisher, but note the black-and-white coloration, especially the white forehead. Typically perches high on exposed branches at forest edges and in clearings with scattered trees, often remaining perched for some time. Nests in large termitaries. The rarely given vocalization is a weak, high-pitched twitter. Widespread, though fairly uncommon in lowlands; to 2,000 ft (600 m). Range: S Mexico to Amazonian Brazil.

Lanceolated Monklet
Micromonacha lanceolata

5 in (13 cm). The brown upperparts, boldly streaked underparts, and pale buffy lores distinguish this diminutive bird. Easily overlooked, given its small size and stoic behavior. Perches fairly low in forests and at forest edge, usually near streams. Gives a series of high-pitched, emphatic, rising notes that increases slightly in pace in the middle portion of the phrase. Rare in Caribbean foothills and middle elevations; from 1,300 to 4,300 ft (400 to 1,300 m). Range: CR to N Bolivia.

Jacamars (GALBULIDAE). With a long, pointed bill and metallic green coloration, the prototypical jacamar bears a curious resemblance to a hummingbird. However, the jacamars closest relatives are the puffbirds (pp. 128-129), with which they share a neotropical distribution, sit-and-wait predator behavior, and the habit of excavating nest burrows in termitaries or the earth—though jacamars are more likely to use vertical embankments. Jacamars also tend to specialize on flying insects, especially butterflies, dragonflies, bees, and wasps; their long bills no doubt help in keeping both the large wings and painful stings of their prey away from their bodies. World: 18, CR: 2

Rufous-tailed Jacamar

Galbula ruficauda

male

female

9 in (23 cm). Once you realize that you're not looking at a hummingbird, identification is straightforward. The male has a white throat; the female's is pale rufous. Usually holds its long bill pointing slightly upward as it actively surveys its surroundings. Prefers forest edges and openings, often on a horizontal perch near eye level. The call note is a sharp *peep!* with a squeaky quality. The song is a long series of squeaks that increases in pitch and pace. Fairly common in wet forest habitats; to 3,900 ft (1,200 m). Range: SE Mexico to NE Argentina.

New World Barbets (CAPITONIDAE). Sometimes considered relatives of the Old World barbets or included in the toucan family, the two groups of New World barbets have recently been separated into their own families (this and the Toucan-Barbets [p. 132]). The term *barbet* comes from the hairlike feathers around the base of the bill, but these are not so prominent in the American barbets. The characteristically thick bill is quite wide at the base. Fruit makes up the bulk of their diet, but they also eat insects. They are sexually dichromatic, with the males being more colorful. All are cavity nesters that generally excavate their own nests in rotting snags. These holes are also used as roosts. World: 14, CR: 1

Red-headed Barbet

Eubucco bourcierii

male

female

6 in (15 cm). The stout, yellow bill and striking head colors of both the male and the female should prevent confusion. Despite the bold color patterns, both sexes can easily go unnoticed as they forage at middle and upper levels of wet forests, often accompanying mixed flocks and rummaging in clusters of dead leaves for insects. Rarely vocalizes. Uncommon at middle elevations from Tilarán Cordillera south; from 1,300 to 5,900 ft (400 to 1,800 m). Range: CR to N Peru.

Toucan-Barbets (SEMNORNITHIDAE). This family comprises just two species, each with limited geographic distribution, one in the highlands of CR and W Panama, the other in the Andes of W Colombia and NW Ecuador. Though not easy to see in the field, the thick bill has a slight hook at the tip of the upper mandible and a two-pronged tip to the lower mandible, into which the upper mandible hook fits. Largely frugivorous, they also take some insects. Very vocal birds, their calls are far carrying. The nest cavity is excavated in a dead snag. Holes in trees are used as communal roosts. World: 2, CR: 1

Prong-billed Barbet

Semnornis frantzii

7 in (17 cm). Stocky, with a stout blue-gray bill and bright ochraceous head and breast, nothing else resembles it. Usually in pairs or small groups, it inhabits epiphyte-laden montane wet forests and adjacent gardens, generally staying well up in the canopy, though coming to within a meter of the ground to feed on fruiting shrubs. The fast-paced, hollow-sounding, even-pitched duets of this species are a characteristic sound of their environment. Common at middle elevations and in highlands; from 2,600 to 7,900 ft (800 to 2,400 m). Range: CR and W Panama.

Toucans (RAMPHASTIDAE). No other bird family is so strongly associated with the neotropics as are the toucans. Their oversized bills enable them to reach ripe fruits hanging at the tips of slender twigs. Though most often seen while feeding at fruiting trees or while calling from exposed perches, all members of the family are also predators that steal both eggs and nestlings from the nests of other species, as well as take insects and small vertebrates as the opportunities arise. Often in small groups, they typically fly in follow-the-leader fashion, each bird separated by several seconds from the next. Their calls consist of either croaking, barking, or yelping sounds. They nest in old woodpecker holes and natural cavities, as toucan bills are unsuitable for the task of excavating. World: 35, CR: 6

Emerald Toucanet

Aulacorhynchus prasinus

12 in (30 cm). The smallest of the CR toucans, it is also the only one with a green head and underparts. Also note the blue on the face and throat, as this characteristic sets the populations of CR and W Panama apart from other subspecies—and may lead to their separation as a distinct species: Blue-throated Toucanet *A. caeruleogularis*. Individuals or pairs forage at middle levels in montane wet forests and adjacent gardens and second growth. Quite vocal; most calls have a harsh, barking quality. Fairly common at middle elevations; from 2,600 to 7,900 ft (800 to 2,400 m). Range: S Mexico to N Bolivia.

Collared Araçari
Pteroglossus torquatus

16 in (41 cm). The dull whitish coloration on much of the upper mandible and the black breast band distinguish this slender toucan from the otherwise similar, but allopatric, Fiery-billed Araçari. Small groups travel through mature wet forests, forest edges, and adjacent gardens. Their flight is direct (not undulating, as in the larger toucans), and they hop jauntily along branches. The most often heard vocalization is a high-pitched, two-note "hiccup." Common in Caribbean lowlands, and decreasingly common up into foothills; uncommon in northern Pacific, where usually found in gallery forest; to 3,300 ft (1,000 m). Range: S Mexico to N Peru.

Fiery-billed Araçari
Pteroglossus frantzii

17 in (43 cm). Very closely related to the Collared Araçari, but differentiated by the mostly red-orange upper mandible and the red band across the upper belly. There is no overlap in their ranges. The behavior and vocalizations of both species are nearly identical. In addition to nesting in unused woodpecker holes, both species also roost in these cavities, with up to six birds sleeping in the same hole. Fairly common in southern Pacific; to 3,900 ft (1,200 m). Range: CR and W Panama.

Keel-billed Toucan
Ramphastos sulfuratus

18 in (46 cm). Who would ever notice that the upper mandible is ridged? The alternative common name of Rainbow-billed Toucan seems far more appropiate for this unmistakable bird. Individuals, pairs, or small groups inhabit forested and semi-open areas, flying across openings with an undulating flight. Common call is a dry, rather frog-like croak, repeated at length and usually given from an exposed perch in the canopy. On Caribbean slope, common in lowlands and fairly common in foothills and middle elevations; on Pacific slope, rare in northern lowlands and fairly common at middle elevations (south to the Central Valley); to 4,600 ft (1,400 m). Range: SE Mexico to NW Venezuela.

Black-mandibled Toucan
Ramphastos ambiguus

22 in (56 cm). Although nearly identical to the Keel-billed Toucan in plumage and behavior, its bill color and voice differ considerably. The local race has a decidedly deep-brown, not black, coloration on the lower mandible and basal half of the upper mandible, but has recently been lumped with a SA relative, hence the incongruous common name. Its song, which is completely different from the Keel-billed Toucan's, is a sonorous *tee de, te-de, te-de* and is one of the most common sounds in its range. On Caribbean and southern Pacific slopes, common in lowlands and decreasingly common up into middle elevations; to 3,900 ft (1,200 m). Range: N Honduras to E Peru.

Woodpeckers (PICIDAE). Widespread—though oddly absent from Madagascar, Australia, and New Zealand—and familiar birds, the woodpeckers are well-adapted to the task they are named after. Two forward-facing and two rear-facing toes, together with stiffened tail feathers, enable a woodpecker to maintain its hold on a vertical trunk as it uses its chisel-shaped bill to chip away at the tree. The purpose of this is not only to construct cavities for nesting and roosting, but also for feeding on insects and larvae that are found below the bark. A very long, barbed tongue with sticky secretions aids in the extraction of the prey. Some species also consume fruit and nectar, and some feed predominantly while on the ground. Sexes are generally similar, with males having more red on the head and/or face. In addition to vocalizations, most species have a characteristic drumming rhythm that is used for communication. World: 230, CR: 16

Acorn Woodpecker
Melanerpes formicivorus

male

female

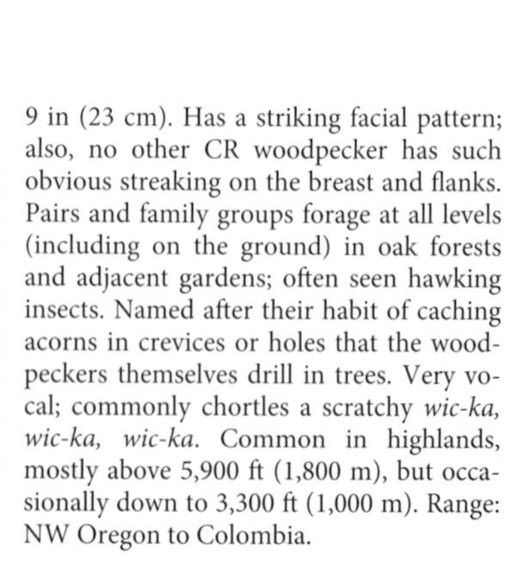

9 in (23 cm). Has a striking facial pattern; also, no other CR woodpecker has such obvious streaking on the breast and flanks. Pairs and family groups forage at all levels (including on the ground) in oak forests and adjacent gardens; often seen hawking insects. Named after their habit of caching acorns in crevices or holes that the woodpeckers themselves drill in trees. Very vocal; commonly chortles a scratchy *wic-ka, wic-ka, wic-ka*. Common in highlands, mostly above 5,900 ft (1,800 m), but occasionally down to 3,300 ft (1,000 m). Range: NW Oregon to Colombia.

Black-cheeked Woodpecker
Melanerpes pucherani

nale

female

⁷ in (18 cm). The only woodpecker on the Caribbean slope with a red belly. (The very similar Gold-
:n-naped Woodpecker *M. chrysauchen* [not illustrated] of the S Pacific has a yellow nape.) Pre-
'ers gardens and forest edges, but also found at openings in mature wet forest, foraging at middle
ınd upper levels; commonly feeds on fruit; drinks nectar from flowers of balsa trees. Makes a fairly
ıigh-pitched, rattling *chirrr, chirrr, chirrr*. Common in Caribbean lowlands and foothills; to 3,000 ft
900 m). Range: S Mexico to W Ecuador.

Hoffmann's Woodpecker
Melanerpes hoffmannii

nale

female

⁷ in (18 cm). This is essentially the only small woodpecker in the NW dry forest. Note the pale under-
ɔarts, black-and-white barred back, and yellow nape. (The very similar Red-crowned Woodpecker
V. *rubricapillus* [not illustrated] of the S Pacific has a red nape.) Prefers gardens, second growth,
leciduous woodlands, and open areas with a few trees; forages at all levels; also feeds on fruit; drinks
ıectar from flowers of balsa trees. Makes a harsh, sputtering rattle. Common in northern Pacific and
n Central Valley, east to Turrialba; to 6,600 ft (2,000 m); uncommon in northern central Caribbean
owlands and foothills. Range: S Honduras to CR.

Rufous-winged Woodpecker
Piculus simplex

male

female

7 in (18 cm). The olive cheeks, pale blue iris, and absence of gray on the crown distinguish it from the similar Golden-olive Woodpecker (which usually is found at somewhat higher elevations). The rufous in the flight feathers is generally not obvious on perched birds. Forages at middle and upper levels of wet forest, and at lower and middle levels in adjacent gardens; often accompanies mixed species flocks. Call is a vigorous, jaylike *cheea, cheea, cheea, cheea*. Uncommon in Caribbean lowlands and foothills, and rare in southern Pacific; to 3,000 ft (900 m). Range: Honduras to W Panama.

Golden-olive Woodpecker
Colaptes rubiginosus

male

female

8 in (20 cm). The pale cheeks, dark iris, and gray crown set it apart from the similar Rufous-winged Woodpecker (typically of lower elevations). Individuals or pairs forage at middle and upper levels of forest; often at lower levels in adjacent second growth and gardens; sometimes accompanies mixed flocks. Utters a rising, sputtering rattle. Fairly common at middle elevations, from 2,300 to 6,600 ft (700 to 2,000 m); also occurs in Caño Negro region. Range: E Mexico to NW Argentina.

Cinnamon Woodpecker
Celeus loricatus

ıale

female

in (20 cm). This and the following species are handsome, crested woodpeckers that are best identi-ıed by determining whether the underparts are a lighter shade than the head and back (this species) r if the head is lighter colored than the body (Chestnut-colored Woodpecker). Found in forest can-py; drops to lower levels at forest edges and in adjacent second growth. Call is a forceful, descend-ng *dwee, dwee, dwe, dwit.* Fairly common in lowlands and foothills of Caribbean slope; to 3,000 ft 900 m). Range: E Nicaragua to SW Ecuador.

Chestnut-colored Woodpecker
Celeus castaneus

ıale

female

in (23 cm). Similar to the Cinnamon Woodpecker (see for ID tips) and like that species feeds redominantly on ants and termites. Forages at middle and upper levels of forest, forest edges, and ıearby gardens; sometimes accompanies mixed species flocks. Like many woodpecker species, it is ften detected by hearing the sound of tapping on wood. Whistles a falling *skeew* succeeded by a ıasal *keh, keh, keh.* Fairly common in Caribbean lowlands and uncommon in adjacent foothills; to ,300 ft (700 m). Range: S Mexico to N Bolivia.

Lineated Woodpecker
Dryocopus lineatus

male

female

13 in (33 cm). This large, dark-backed woodpecker is similar to the Pale-billed Woodpecker, but i differentiated by the whitish stripe running from the base of the bill down the side of the neck; als note the dark bill. Generally prefers more open habitat than the Pale-billed Woodpecker; forage at all levels. Gives an emphatic *wick-wick-wick-wick-...*; the drum is very rapid, lasting about thre seconds. Fairly common from lowlands to middle elevations; to 4,600 ft (1,400 m). Range: Mexic to NE Argentina.

Pale-billed Woodpecker
Campephilus guatemalensis

male

female

13 in (33 cm). In addition to its pale bill, this large woodpecker is told apart from the similar Lin eated Woodpecker by the lack of striping on the face (having an essentially entirely red head). Mor commonly found in forest habitat than the Lineated Woodpecker; forages at all levels of forests occasionally visits forest edges and adjacent gardens. Its characteristic sound is a loud, sharp, doubl rap; its call is a sputtering, laughing rattle. Common in lowlands, uncommon in foothills and middl elevations; to 4,600 ft (1,400 m). Range: Mexico to W Panama.

Ovenbirds, Woodcreepers (FURNARIIDAE). Brown and rufous are the predominant colors in this neotropical family of diverse habits and habitats. Woodcreepers forage by climbing up the trunks and branches of trees in search of insects, but typically do not peck into wood the way woodpeckers do. They are most common in lowland wet forests, where tree trunks are not so covered with mosses and other epiphytes. The remaining members of the family typically forage for invertebrates by a variety of methods that include rummaging through clusters of dead leaves, gleaning from fresh vegetation, pecking into dead twigs and vines, and poking about in epiphytes. Many, including the woodcreepers, are cavity nesters; others construct nests that, in some species, are quite elaborate structures. World: 298, CR: 34

Wedge-billed Woodcreeper

Glyphorynchus spirurus

6 in (15 cm). Lacking any striking pattern, this small woodcreeper can be recognized by its short, slightly upturned bill. Forages at middle and lower levels of mature wet forest and advanced second growth; often accompanies mixed flocks and also attends army-ant swarms. Sings a quick *wididid-ididi* that falls then rises. Common from lowlands to middle elevations on Caribbean and southern Pacific slopes; to 4,900 ft (1,500 m). Range: S Mexico to N Bolivia.

Northern Barred-Woodcreeper

Dendrocolaptes sanctithomae

11 in (28 cm). No other large woodcreeper has an entirely dark bill; the fine black barring on the head, back, and underparts is also distinctive. Forages at middle and lower levels of mature forest, advanced second growth, and adjacent gardens; often accompanies army-ant swarms. Calls at dawn and dusk with a penetrating *dooowee, dooowee, dooowee, dooowee,* each part upslurred. On both slopes, common in wet lowlands and uncommon in foothills, to 4,300 ft (1,300 m); rare in dry lowlands. Range: SW Mexico to NW Venezuela.

Cocoa Woodcreeper

Xiphorhynchus susurrans

9 in (23 cm). Very similar to several other woodcreepers with streaked heads, but distinguished by its fairly stout and straight bicolored bill (note the dark upper mandible and pale lower mandible). Forages at all levels of forest edges, second growth, and gardens; occasionally found in mature forest. Very vocal, but typically calls hidden from view; gives a series of seven or more emphatic notes (*wheet, wheet, wheet ...*), rising slightly then falling off at the end. Common in wet lowlands and foothills, to 3,000 ft (900 m); rare in dry lowlands. Range: E Guatemala to NE Venezuela.

Streak-headed Woodcreeper
Lepidocolaptes souleyetii

8 in (20 cm). Very similar to several other woodcreepers with streaked foreparts, but note the fairly slender, slightly decurved, pinkish-brown bill. Forages at all levels in garden habitats. Gives a two- or three-second, descending trill. Common and widespread in lowlands, uncommon in foothills and middle elevations; to 4,900 ft (1,500 m). Range: W Mexico to NW Peru.

Spot-crowned Woodcreeper
Lepidocolaptes affinis

8 in (20 cm). This is the only common woodcreeper at higher elevations (almost no range overlap with similar Streak-headed Woodcreeper). Has small buffy spots on brown crown; slender, slightly decurved, grayish bill is darker at base and paler at tip. Forages at all levels, both in garden habitats and mature oak forests; often accompanies mixed species flocks. Gives one or two high, nasal squeaks followed by a fairly even two or three-second trill that drops off and slows slightly at the very end. Fairly common in Talamanca Cordillera and uncommon in Central Cordillera, from timberline down to 4,900 ft (1,500 m), rarely as low as 3,300 ft (1,000 m). Range: NE Mexico to W Panama.

Plain Xenops
Xenops minutus

5 in (13 cm). Identified by the combination of a slightly upturned bill, buffy superciliary, and whi malar stripe. Forages at all levels in mature wet forest, forest edges, second growth, and garden behaves like a miniature woodpecker, pecking into vines and dead twigs. Accompanies mixed flock Gives a fast, high-pitched trill. Fairly common in lowlands and foothills of southern Pacific, rare i northern Pacific; on Caribbean slope, uncommon in lowlands and foothills; to 4,900 ft (1,500 m Range: S Mexico to NE Argentina.

Ruddy Treerunner
Margarornis rubiginosus

6 in (15 cm). No other CR furnariid is so uniformly reddish-brown with whitish throat and superciliary. Almost always encountered accompanying mixed species flocks, typically with Sooty-capped Chlorospingus (p. 230). Climbs trunks and hangs upside-down below limbs as it forages in moss and epiphytes at all levels of wet montane forest, forest edges, and adjacent gardens. Rapidly twitters high, thin notes. Common in highlands; from 4,900 ft (1,500 m) to timberline. Range: CR to W Panama.

Red-faced Spinetail
Cranioleuca erythrops

6 in (15 cm). The combination of an olive-brown body with rufous wings, tail, and half hood is unique. Forages acrobatically on mossy limbs of mature wet forest and at forest edges, mostly at middle levels; often rummages in dead leaf clusters; accompanies mixed flocks. Builds a large, conspicuous nest of moss and fibers at the end of a slender, hanging branch in the mid-canopy. Delivers a high, rapid, slightly descending series of staccato notes. Fairly common at middle elevations; from 2,300 to 6,600 ft (700 to 2,000 m). Range: CR to W Ecuador.

Typical Antbirds (THAMNOPHILIDAE). To seasoned birders, this family epitomizes the lowland neotropic rainforests (only a few species prefer other habitats). Though generally lacking in bright coloration, many antbirds are handsomely patterned—but seeing them well in the dim understory is part of the challenge. Fortunately, most tend to be quite vocal, which helps in tracking them down. Despite the "ant" prefix, none routinely feed on ants. The name derives from several species' habit of following army ant swarms in order to feed on the arthropods that flee the ants. Witnessing this phenomenon is truly a highlight of any birding trip. Antbirds are typically monogamous and both members of the pair assist with breeding chores. World: 232, CR: 22

Black-hooded Antshrike

Thamnophilus bridgesi

male

female

6 in (15 cm). The small white spots on the shoulder identify the otherwise black male. The female has white streaking on the head and underparts, but is mostly olive-gray. Pairs inhabit lower levels of mature forest, second growth, and forest edges; often with mixed species flocks. Gives an accelerating series of staccato notes that ends in a longer, lower note. Common in southern Pacific; rare north of Carara NP, to Tenorio Volcano; to 3,900 ft (1,200 m). Range: CR and W Panama.

nale

Fasciated Antshrike
Cymbilaimus lineatus

'emale

7 in (18 cm). Male told apart from male Barred Antshrike by red iris and black crown. Female is unique in being barred with black and buff; also note rufous crown. Pairs hop methodically through tangled vegetation at lower and middle levels of forest edge and second growth. Delivers four to eight clear notes (*kwew, kwew, kwew, kwew*) at the same pitch, but each note rises at the end. Fairly common in wet Caribbean lowlands, rare in Caño Negro region and foothills; to 3,900 ft (1,200 m). Range: SE Honduras to Amazonian Brazil.

nale

Barred Antshrike
Thamnophilus doliatus

female

6 in (15 cm). The male's combination of a pale iris and black-and-white barred plumage is unique. The female is the only cinnamon-rufous bird in CR with streaking on the side of the head. Pairs typically stay low in thickets, forest edges, and second growth. Call is a fast-paced series of chuckling notes that accelerates before ending in a higher-pitched, nasal *wek!*. Common in northern Pacific and Caño Negro region, uncommon in wet lowlands and foothills; to 4,600 ft (1,400 m). Range: E Mexico to N Argentina.

Dot-winged Antwren
Microrhopias quixensis

male

female

4 in (10 cm). On both sexes, note the single white wing bar, white spots on shoulder, and white tips to tail feathers. Pairs and small groups forage actively with mixed flocks in lower levels of mature wet forest and adjacent advanced second growth. Sings a series of high, fast notes, which rise initially in pitch and intensity before descending; also gives a sharp *speEa*. Common in southern Pacific lowlands and foothills, uncommon in Caribbean lowlands and foothills; to 3,300 ft (1,000 m). Range: SE Mexico to N Bolivia.

Chestnut-backed Antbird
Myrmeciza exsul

male

female, Pacific race

6 in (15 cm). No other bird with blue orbital skin combines a slaty head and a chestnut back. Pairs forage near (or even on) the ground in mature wet forest. Whistles two or three clear, labored notes, the last one lower (reminiscent of Black-faced Antthrush [p. 152] and Black-throated Trogon [p. 121]); also gives a nasal, raspy *hyaah*. Common in wet lowlands and foothills of southern Pacific, but rare north of Carara NP; fairly common in Caribbean lowlands and foothills; to 3,900 ft (1,200 m). Range: E Nicaragua to W Ecuador.

Bicolored Antbird
Gymnopithys bicolor

6 in (15 cm). The combination of blue orbital skin, brown upperparts, and mostly white underparts is distinctive. Pairs or small groups forage near the ground in mature wet forest and adjacent advanced second growth, almost exclusively at army-ant swarms. Frequently utters a harsh *nhyarr*; sings a series of high notes that increase in pitch and pace, then slow down and descend (similar to song of Ocellated Antbird [p. 150]). Uncommon in wet lowlands and foothills of Caribbean and southern Pacific slopes, to 5,200 ft (1,600 m); rare on Pacific slope of Guanacaste Cordillera. Range: E Honduras to NE Peru.

Spotted Antbird
Hylophylax naevioides

male

female

4 in (10 cm). The dapper male is boldly patterned and quite distinctive. The female is browner and has a muted pattern. Individuals and pairs forage in the understory of mature wet and humid forests and adjacent advanced second growth, very often with army ants. Sings a soft, sweet, buzzy, descending series of eight to ten doubled notes: *zpeeda, zpeeda, zpeeda...* .Common on both slopes of Guanacaste Cordillera, from 2,000 to 2,600 ft (600 to 800 m); uncommon elsewhere in Caribbean lowlands and foothills, to 3,300 ft (1,000 m). Range: E Honduras to W Ecuador.

Ocellated Antbird
Phaenostictus mcleannani

8 in (20 cm). The large area of sky-blue orbital skin and the pronounced scalloping on the belly and back set this species apart. Pairs or small groups forage near the ground in mature wet forest and adjacent advanced second growth, almost exclusively at army-ant swarms. Song quite similar to that of Bicolored Antbird (p. 149), but more dulcet; also gives a descending, buzzy *dzjeer*. Uncommon in Caribbean lowlands and foothills; to 3,900 ft (1,200 m). Range: E Honduras to NW Ecuador.

Antpittas (GRALLARIIDAE). The antpittas are characterized by plump bodies, big eyes, long legs, and short tails. They stay low in forest and tangled second-growth, where their earthen tones provide cryptic protection and trying to view them can be exceedingly frustrating. Some species will hop about on trails in the dim light of dawn and dusk. Though they feed primarily on arthropods, they do not regularly attend army-ant swarms. Relatively little is known of their nesting, but it seems that pairs share in the duties. World: 53, CR: 4

Streak-chested Antpitta
Hylopezus perspicillatus

5 in (13 cm). The most readily seen antpitta in CR; its odd proportions, together with a buffy eye ring, dark malar stripe, streaked breast, and wing bars are diagnostic. Prefers relatively open areas in understory of mature wet forest; habitually inflates and deflates abdomen. Its song (of up to ten notes) lasts four seconds. The first three notes rise slightly; the following notes are even pitched and slower; the last few notes are coupled. Fairly common in southern Pacific, uncommon on Caribbean slope; to 3,900 ft (1,200 m). Range: NE Honduras to NW Ecuador.

Antthrushes (FORMICARIIDAE). Like their relatives the tapaculos and antpittas, the antthrushes are dark birds of the forest understory that are heard far more often than seen. They feed on arthropods and will opportunistically forage at an army-ant swarm when one passes through their territory, but usually are found singly. Known to nest in hollow stumps and palm trunks, typically not more than ten feet above the ground. World: 11, CR: 3

Black-faced Antthrush
Formicarius analis

7 in (18 cm). Gives the impression of a small, brown chicken as it struts with its tail cocked up. Forages in leaf litter of mature forest and advanced second growth. Whistles a labored *pyee, pyew, pyew*; the first note (accented and higher) is usually followed by two or three notes, but sometimes by as many as ten or more (compare with call of Chestnut-backed Antbird [p. 149]). Common in wet lowlands and foothills; to 1,600 ft (500 m) on Caribbean slope (higher in the south) and to 4,900 ft (1,500 m) in southern Pacific. Range: S Mexico to N Bolivia.

Tapaculos (RHINOCRYPTIDAE). The single CR species is the northernmost representative of this essentially South American family of basically terrestrial birds. Many members of the genus *Scytalopus* are so similar in appearance that they can only be told apart by voice. Seeing them is extraordinarily difficult anyway, given their mouse-like habits. Insects comprise the bulk of the diet, but some species are also known to consume seeds and fruit. It seems that both members of a pair assist in nesting activities. World: 57, CR: 1

nale

Silvery-fronted Tapaculo

Scytalopus argentifrons

'emale

4 in (10 cm). A small dark bird that blends in well with the forest floor. The male shows a whitish stripe across the brow; the brownish female is essentially devoid of field marks, though both sexes have barred flanks. Inhabits dense undergrowth in wet montane forests, especially along streams and ravines. At intervals, utters a characteristic series of rapid, emphatic, staccato notes that increase in volume, then decrease in pace, lasting five seconds or more. Common from timberline down to 4,900 ft (1,500 m) on Pacific slope, and down to 3,900 ft (1,200 m) on Caribbean slope, from Miravalles Volcano south. Range: CR and W Panama.

Tyrant Flycatchers (TYRANNIDAE). The New World flycatchers constitute the world's largest family of birds and, as such, demonstrate much variety in form, behavior, and habitat choice. The archetypical flycatcher is rather drably attired and perches in the open, where it flies out to capture passing insects. However, numerous species sport bright colors and/or boldly contrasting patterns, and many use other foraging techniques. In fact, most tropical species consume good quantities of fruit, in addition to arthropods. Virtually every habitat from mangroves to main streets of populous cities is home to some sort of flycatcher. Nest architechture is likewise diverse, varying from minimalist platforms of sticks to nests that, in relation to the size of the builders, are some of the largest in the world. World: 419, CR: 82

Yellow-bellied Elaenia

Elaenia flavogaster

6 in (15 cm). Readily identified when crest is fully raised, as it is parted down the middle; otherwise voice is the best clue with this rather generic flycatcher. Found in semi-open habitats and gardens where it forages at middle levels, frequently eating berries. Often appears agitated due to the raised crest and somewhat irritated quality of the freely given call, which is a loud, wheezy *wheeeur*. Widespread and fairly common in most of CR, though rare in northern Pacific; to 7,200 ft (2,200 m). Range: S Mexico to Argentina.

Mountain Elaenia

Elaenia frantzii

6 in (15 cm). Has a round head. The wing bars often give the appearance of being spotted on this otherwise drab-olive bird. Found at forest edges and in gardens; eats mostly berries. Gives a loud whistled *pseer*. Common in highlands, from about 3,900 ft (1,200 m) to timberline. Range: Guatemala to N Venezuela.

Torrent Tyrannulet
Serpophaga cinerea

4 in (10 cm). The combination of blackish cap, wings, and tail on a small gray bird, together with habitat, distinguish it (compare with the sympatric American Dipper [p. 188]). Forages from river rocks and along banks, often pumps tail. Call is a shrill *tseep*. A common inhabitant along swift, rocky, highland streams, from about 1,600 to 7,200 ft (500 to 2,200 m)—occasionally below this range on Caribbean slope—from Tilarán Cordillera south. Range: CR to W Bolivia.

Ochre-bellied Flycatcher
Mionectes oleagineus

5 in (13 cm). Named after its only field mark, this is one of those species where the absence of field marks is the key identifying feature. Stays low inside mature forest and second growth; also comes to forest edges and gardens. Often flicks one wing up over the back, then the other. Males sing incessantly from low perches, producing a series of sharp *chip* notes that is followed by a number of louder, more emphatic notes. Common in wet lowlands of both slopes, uncommon in foothills and in Central Valley, and rare in gallery forests of northern Pacific lowlands; to 4,600 ft (1,400 m). Range: S Mexico to N Bolivia.

Paltry Tyrannulet
Zimmerius vilissimus

4 in (10 cm). On this small flycatcher, note the yellow wing edging, the whitish line from across the brow to just behind the eye, and the short, dark bill. Frequents gardens, forest edges, and second growth, dining on mistletoe and other berries, as well as insects. Call is a whistled *peee-yer* that falls slightly at the end. Common and widespread, from lowlands to timberline, though rare in Central Valley and northern Pacific. Range: S Mexico to N Venezuela.

Scale-crested Pygmy-Tyrant
Lophotriccus pileatus

3 in (8 cm). Though seldom raised, the rufous-tipped black crest feathers set this tiny bird apart from all other drab-olive flycatchers. Prefers middle levels of wet forests and tall second growth. Produces an assertive series of dry, wooden *chik* notes. Common on Caribbean slope, from 1,000 to 5,600 ft (300 to 1,700 m); common on Pacific slope, from 2,500 to 5,600 ft (750 to 1,700 m). Also found in hills of Osa Peninsula. Range: CR to SE Peru.

Northern Bentbill
Oncostoma cinereigulare

4 in (10 cm). The curved bill is the diagnostic feature; also note yellow wing edging and pale iris. Although it typically forages from one to six meters above the ground in second growth and more open areas of forests, its small size, drab color, and staid manner, combined with darting flight, make it difficult to see. The often heard, soft, rolling *perrrrrr* alerts one to its presence. Common in southern Pacific lowlands and foothills, uncommon in northern Pacific and Caribbean lowlands, also uncommon in foothills; occasionally found as high as 3,600 ft (1,100 m). Range: S Mexico to W Panama.

Common Tody-Flycatcher
Todirostrum cinereum

ventral view

dorsal view

4 in (10 cm). The yellow iris imparts a glaring look, made all the more menacing by the proportions of the bill on this small and colorful flycatcher. Forages actively in lower levels of vegetation in gardens, second growth, and along streams. Calls with high, staccato *chip* notes, which often run together very rapidly. Common and widespread, though less common in northern Pacific; to 5,200 ft (1,600 m). Range: S Mexico to N Paraguay.

Yellow-olive Flycatcher
Tolmomyias sulphurescens

5 in (13 cm). Another generic flycatcher, it can be distinguished by the pale iris and the wide base of the bill. Forages at lower and middle levels of second growth, forest edges, and gardens. The conspicuous nest, in the form of an upside-down U, is made of fine black fibers. Call is a shrill, sibilant *tssssp*. Common and widespread; to 4,600 ft (1,400 m). Range: S Mexico to N Argentina.

White-throated Spadebill
Platyrinchus mystaceus

4 in (10 cm). The pied facial pattern and exaggerately wide bill, together with the overall brownish coloration, distinguish it from all but the very similar Stub-tailed Spadebill *P. cancrominus* (not illustrated) of lower and drier elevations (no range overlap). Tiny and easily overlooked if it sits quietly. Keeps within three meters of forest floor. Call is a sharp, squeaky *squik!* Common in understory of middle-elevation wet forest; from 2,300 to 6,900 ft (700 to 2,100 m). Range: CR to NE Argentina.

Royal Flycatcher
Onychorhynchus coronatus

dorsal view

bird at nest

7 in (18 cm). The amazing crest is rarely opened, though it can sometimes be seen on preening birds. Otherwise, note the bright tawny tail, buffy rump, and long, flat bill. Compare with Bright-rumped Attila (p. 164). Found at lower levels of humid forest, forest edges, and particularly along wooded streams. Constructs a pendant nest nearly two meters long. Call is a resonant, whistled *keyerink*. Uncommon in Pacific lowlands, rare in Caribbean lowlands; to 3,000 ft (900 m). Range: SE Mexico to SE Brazil.

Sulphur-rumped Flycatcher
Myiobius sulphureipygius

5 in (13 cm). Could be mistaken for a Buff-rumped Warbler (p. 207), but note lack of eye line and superciliary. The hint of a dark bar below the eye and the bright ochraceous breast distinguish it from the otherwise similar Black-tailed Flycatcher *M. atricaudus* (not illustrated) of the southern Pacific lowlands and foothills. Forages at lower and middle levels of wet forests, often with mixed species flocks. Call is an explosive *swit!* Fairly common in lowlands and foothills; to 3,900 ft (1,200 m). Range: SE Mexico to W Ecuador.

Tufted Flycatcher
Mitrephanes phaeocercus

5 in (13 cm). The crest and overall coloration distinguish it from all but the larger Ochraceous Pewee, so note the rather short bill. Perches on exposed twigs in treefall gaps and other openings in mature forest, also at forest edges and gardens; often in pairs. Typically shivers its wings and tail upon alighting after a sally, and often gives a quick, spirited series of *pip*'s. Common in middle elevations and highlands; from 1,600 to 9,800 ft (500 to 3,000 m). Range: Mexico to NW Ecuador.

Ochraceous Pewee
Contopus ochraceus

7 in (18 cm). Larger than the very similar Tufted Flycatcher and with a proportionately longer bill. Sallies from exposed perches in treefall gaps and other openings in oak forest; shakes tail upon landing. Call is a shrill *pit, pit, pit*. Rare in Talamanca Cordillera and on Irazú and Turrialba Volcanoes; from 7,200 to 9,800 ft (2,200 to 3,000 m). Range: CR and W Panama.

Tropical Pewee
Contopus cinereus

ventral view

dorsal view

5 in (13 cm). Quite similar to the migrant Eastern Wood-Pewee and Western Wood-Pewee *Contopus sordidulus* (not illustrated), but has pale lores and the crown is a shade darker than the nape and back. Frequents gardens, pastures with brush, and forest edges; tends to perch quite low and does not regularly return to the same perch after a sally. Call is a rapid, bubbly, trill. Common in Caribbean lowlands and foothills; to 3,300 ft (1,000 m). Uncommon in Central Valley and in southern Pacific, up to 4,300 ft (1,300 m); also uncommon in mangroves around Gulf of Nicoya. Range: SE Mexico to NE Argentina.

Eastern Wood-Pewee
Contopus virens

6 in (15 cm). The orange lower mandible with dark tip sets it apart from the Western Wood-Pewee *C. sordidulus* (not illustrated), which only shows a hint of orange at the base of the lower mandible. Occurs in virtually any habitat where there is an exposed perch. Birds habitually return to the same perch after a sally. Call is a rising *pee-wee*; sometimes whistles its full *pee-a-weee*. A common and widespread passage migrant from mid-Aug to late Nov and from mid-March to mid-May; to 4,900 ft (1,500 m), and rarely to 8,200 ft (2,500 m). Rare winter resident; to 3,900 ft (1,200 m). Range: Breeds in E NA, winters to N Bolivia and W Brazil.

Yellow-bellied Flycatcher
Empidonax flaviventris

5 in (13 cm). The combination of a rounded head, an eye ring, and a dull-yellow throat helps to tell it apart from any similar-looking flycatcher. Inhabits lower levels of forest and second growth. Frequently whistles a clear, rising *tsu-wee*, reminiscent of a wood-pewee. Common from lowlands to middle elevations; to 4,900 ft (1,500 m), from late Aug to late May. Range: Breeds in E NA, winters south to Panama.

Yellowish Flycatcher
Empidonax flavescens

5 in (13 cm). The eliptical eye ring and buffy-brown wing bars are distinctive features of this often fairly tame flycatcher. Inhabits lower and middle levels of mature wet forest and adjacent second growth and gardens. Call is a single, high *tseew*. Common in middle elevations and highlands; from 2,600 to 8,200 ft (800 to 2,500 m). Range: S Mexico to N Panama.

Black-capped Flycatcher
Empidonax atriceps

4 in (10 cm). The eliptical eye ring and dark crown are definitive on this small, highland flycatcher. Forages at all levels of forest edge, brushy pastures, and gardens; often quite low and confiding. Call is a single *pwip*. Common in highlands, from about 7,200 ft (2,200 m) to above timberline. Range: CR and W Panama.

Black Phoebe
Sayornis nigricans

7 in (18 cm). No other dark, sooty bird in CR combines black bill and eye with white lower belly. Almost always near water; often perches on river rocks, but also on wires, buildings, and the ground. Gives a quick, shrill *chirp*. Common in Central Valley and at middle elevations, uncommon on Pacific slope of Guanacaste and Tilarán Cordilleras; from 1,600 to 7,200 ft (500 to 2,200 m). Range: W US to NW Argentina.

Long-tailed Tyrant
Colonia colonus

10 in (25 cm). Elongated central tail feathers (that contribute nearly half its total length) are diagnostic; also note whitish superciliary and back. Forages from exposed perches in forest clearings and gardens, and along rivers; often in pairs. Nests in old woodpecker holes. Call is a sweet, rising *sweeE*. Fairly common in Caribbean lowlands, to 2,000 ft (600 m). Range: SE Honduras to NE Argentina.

Bright-rumped Attila
Attila spadiceus

dorsal view

ventral view

8 in (20 cm). If seen, the yellowish rump is quite distinctive. Otherwise, note the red iris, hook-tipped bill, faint superciliary, and diffuse streaking on breast. Found in mature forests, second growth, forest edges, and gardens. Very vocal (though difficult to see when calling); gives an increasingly emphatic *wit, wi-deet, wi-deet, wi-deet, wheeew*, the last note dropping down; also makes a fast *we-hir,we-hir,we-hir,we-hir, we-hir,we-hir* that rises and falls. Common and widespread; to 5,900 ft (1,800 m). Range: NW Mexico to SE Brazil.

Rufous Mourner
Rhytipterna holerythra

8 in (20 cm). This nondescript rufous bird is best identified by its vocalizations. Whistles a loud, slow, somewhat sad *wheea-tseer* that rises and falls. Mostly found inside mature forest and tall second growth, at middle to upper levels; often accompanies mixed species flocks. Fairly common in wet lowlands and foothills; to 3,900 ft (1,200 m). Range: SE Mexico to NW Ecuador.

Dusky-capped Flycatcher
Myiarchus tuberculifer

6 in (15 cm). Crown that is darker than the nape and back, along with an all-dark bill, serves to tell it apart from other *Myiarchus* flycathers. Members of this genus are cavity nesters. Forages at low and medium heights in gardens and at forest edges. Whistles a somewhat mournful *wheeew* that rises and falls slightly. Common and widespread; to 5,900 ft (1,800 m). Range: SW US to NW Argentina.

Brown-crested Flycatcher
Myiarchus tyrannulus

8 in (20 cm). Similar to other *Myiarchus* flycathers, but note decidedly brownish crown and all-dark bill. (The common migrant Great-crested Flycatcher *M. crinitus* [not illustrated] has obviously pale basal half of lower mandible.) Favors open habitats—such as brushy pastures—and forest and mangrove edges. Call is a repetitious, sharp *whip*. Common in northern Pacific and uncommon in western end of Central Valley; to 3,000 ft (900 m). Range: SW US to NE Argentina.

Great Kiskadee
Pitangus sulphuratus

9 in (23 cm). Behavior and coloration combine to make this one of the most obvious birds in CR. Distinguished from other large yellow-bellied flycatchers by combination of white superciliaries that form a complete ring across the nape and bright rufous edging on wing and tail feathers. Frequently displays concealed yellow crown patch. Perches openly in gardens and clearings; builds a bulky straw nest, often situated on utility poles. A very vocal bird, its English common name is an onomatopoeic version of the loud three-note call: *kis-ka-dee* (somewhat similar to call of Gray-capped Flycatcher [p. 168]). Common and widespread; to 5,900 ft (1,800 m). Range: S Texas to central Argentina.

With yellow crown patch exposed.

Boat-billed Flycatcher
Megarynchus pitangua

9 in (23 cm). The exaggeratedly large bill sets it apart from the very similar Great Kiskadee; also note that the superciliaries do not quite meet on the nape. Frequents gardens and forest edges, where it seems to specialize on rather large insects (e.g., cicadas) taken in flight from a substrate, rather than in mid-air. Produces a rattling call reminiscent of most CR *Melanerpes* woodpeckers (pp. 136-137). Fairly common and widespread; to 7,200 ft (2,200 m). Range: Mexico to N Argentina.

Social Flycatcher
Myiozetetes similis

6 in (15 cm). Note the disproportionately small bill on this fairly large, yellow-bellied flycatcher, and that the superciliaries do not quite meet on the nape. In addition to insects, eats small fruits. Found at lower and middle levels in gardens and along streams. Calls with an agitated *chip, cheer-de-cheer-de, chip* that alternates up and down (song of birds in southern Pacific has a squeakier and less colorful quality). Very common and widespread; to 7,200 ft (2,200 m). Range: Mexico to NE Argentina.

Gray-capped Flycatcher
Myiozetetes granadensis

6 in (15 cm). Though it is the same size and shape of Social Flycatcher (p. 167), perhaps more likely confused with Tropical Kingbird (p. 170), but note white stripe across brow that extends to behind the eye. Forages at lower and middle levels in gardens and along streams. In addition to insects, consumes small fruits. Call is a single, short *wick*; also makes an agitated, squeaky *bee-be-dee*, recalling a Great Kiskadee (p. 166). Common in wet lowlands, fairly common in foothills and middle elevations; to 4,900 ft (1,500 m). Range: E Honduras to N Bolivia.

Streaked Flycatcher
Myiodynastes maculatus

8 in (20 cm). Resembles Sulphur-bellied Flycatcher, but has pale basal half of lower mandible, whitish chin and underparts, and pale golden superciliary. Prefers middle and upper levels of gardens and forest edges. Call is a harsh, squeaky *pick-pyew*. On Pacific slope, fairly common in lowlands and uncommon in foothills; to 3,900 ft (1,200 m). Casual on Caribbean slope. Range: E Mexico to central Argentina.

Sulphur-bellied Flycatcher
Myiodynastes luteiventris

dorsal view

ventral view

8 in (20 cm). Distinguished from Streaked Flycatcher by all-dark bill, black chin and malar stripe, whitish superciliary, and decidedly yellowish wash to underparts. A noisy inhabitant of gardens and forest edge, typically at upper levels. Call is a high, squeaky *swee-eeah*. Common passage migrant, from late Feb to mid-May and from early Aug to mid-Oct; common breeding migrant from April to Sept. Passage migrants can show up almost anywhere; to 7,200 ft (2,200 m). Breeding migrants nest in northern half of CR, including the Central Valley; to 6,600 ft (2,000 m); do not nest in Caribbean lowlands. Range: Breeds from SE Arizona to Costa Rica, winters from E Ecuador to N Bolivia.

Piratic Flycatcher
Legatus leucophaius

6 in (15 cm). The head pattern recalls various other flycatchers, but note the combination of small bill and faintly streaked breast. Favors gardens and forest edges, typically perches in treetops. Named for its habit of expropriating other birds' nests by pestering the builders until they abandon the structure. Often heard call is a high, ringing *bee-ee* that rises and falls, followed after a pause by a series of three to six fast notes: *bidididi*. Common breeding resident from lowlands to about 1,500 m; migrates to SA after breeding (most gone by late Sept), begins returning in late Jan. Range: SE Mexico to N Argentina.

Tropical Kingbird
Tyrannus melancholicus

8 in (20 cm). Of all of the large, yellow-bellied flycatchers in CR, it has the plainest head, with just a slightly darker wash through the eye. Also note the notched tail, visible when perched, that distinguishes it from the very similar Western Kingbird *T. verticalis* (not illustrated), an uncommon winter resident in the dry northwestern part of the country. Usually in pairs, in open areas with at least some trees. Perches in exposed situations, particularly powerlines, which are vantage points for spotting flying insects that are chased down in aerial pursuit. It also readily eats berries. Sings a lively *bee-bibididi*. Very common; to 7,500 ft (2,300 m). Range: S Arizona to central Argentina.

Scissor-tailed Flycatcher
Tyrannus forficatus

13 in (33 cm), including the elongated outer tail feathers. The extremely long tail and salmon flanks are diagnostic. Prefers open areas with scattered trees, but also perches on fences and wires. Gives a flat *pik*. A common winter resident in northern Pacific lowlands and foothills, uncommon in Central Valley (where occasionally found above 6,600 ft [2,000 m]), rare in southern Pacific lowlands, and casual in northern central Caribbean lowlands; from Oct to April. Range: Breeds in central US, winters south to W Panama.

Fork-tailed Flycatcher

Tyrannus savana

14 in (36 cm), including the elongated outer tail feathers. An exaggerately long tail, along with dark upperparts contrasting with white underparts, sets it apart. Found in open areas, where it often perches very low in trees or on bushes, if not on the ground. Fairly common in southern Pacific intermontane valleys; from 300 to 3,900 ft (100 to 1,200 m). Rare in Cartago area and casual elsewhere (e.g., northern Pacific and Caribbean lowlands). Range: S Mexico to S SA.

Tityras, Becards, Allies (TITYRIDAE). Formerly considered members of Cotingidae or Tyrannidae, the tityras and becards are currently placed in their own family. Principally insectivorous, they also take berries. Tityras are cavity nesters that line their hole with dead leaves, even covering the eggs and nestlings prior to leaving the nest. Becards construct rather large, rounded, untidy looking nests. The female does all or most of the incubation, while both parents feed the nestlings. World: 33, CR: 9

Black-crowned Tityra
Tityra inquisitor

male

female

7 in (18 cm). Similar in plumage to the Masked Tityra, but has entirely dark bill and fully feathered face. Male has black crown; female has chestnut face. Pairs or small groups visit fruiting trees in gardens and at forest edges. Infrequent vocalizations are softer than those of Masked Tityra. Widespread but uncommon; to 3,900 ft (1,200 m). Range: SE Mexico to NE Argentina.

Masked Tityra
Tityra semifasciata

male

female

8 in (20 cm). The red orbital area and basal half of bill on this otherwise mostly white bird is quite distinctive. Pairs or small groups visit fruiting trees in gardens and at forest edges. Frequently makes dry, buzzy squeaks that could suggest an insect noise or a bathtub toy. Common and widespread; to 5,900 ft (1,800 m). Range: Mexico to N Bolivia.

Cinnamon Becard
Pachyramphus cinnamomeus

6 in (15 cm). Told apart from all other cinnamon-rufous species by faint buffy supraloral stripe and dusky lores. Male and female identical. Caribbean-slope birds are found at lower and middle levels of gardens and forest edges; Pacific birds favor vicinity of mangroves. Whistles a high, melancholy, descending *twee, twee-tee-tee-tee.* Common in Caribbean lowlands and foothills; to 3,000 ft (900 m). Uncommon in Pacific coastal lowlands, from Gulf of Nicoya south. Range: SE Mexico to S Ecuador.

Rose-throated Becard

Pachyramphus aglaiae

female

7 in (18 cm). Despite the English common name, the male of the local race does not show any color on the throat and is essentially entirely dark gray (darkest on crown and palest on underparts). The female is the only cinnamon-rufous bird with a blackish cap. Pairs forage at middle and upper levels of mature dry forest, at edges of humid forest, and in gardens. Gives a sharp, upwardly inflected *swee-ah*. Fairly common resident in northern Pacific lowlands, uncommon in western Central Valley and in southern Pacific; to 3,900 ft (1,200 m). Casual migrant in Caribbean lowlands, from Nov to March. Range: SW US to W Panama.

male

Manakins (PIPRIDAE). Other than the incredible Birds-of-Paradise, perhaps no other group of birds has evolved such elaborate courtship displays as the manakins. A staple diet of small fruit, together with year-round availability of their food, has freed the males to devote an inordinate amount of time to their displays. Sexual selection has also driven them to evolve unique feather structures that play a role in their performances. The colorful male has nothing to do with domestic duties, rather it is the drab female that alone handles all of the nesting activity. Many species frequently bathe in forest streams in the late afternoon. World: 54, CR: 9

Long-tailed Manakin
Chiroxiphia linearis

male

juvenile male

Male 10 in (25 cm); female 6 in (15 cm). The greatly elongated central tail feathers, together with a scarlet crown and a powder-blue back, readily identify the male. The olive female has bright orange legs and slightly protruding central tail feathers. Juvenile male resembles female, but has scarlet crown. Forages at lower and middle levels of forest and advanced second growth. Calls include a liquid *who-wee-do* and a nasal *waahh.* Males display on low perches, while emitting an undulating *chur-wu, chur-wu,* with an underlying buzzy chatter. Common in northern Pacific and western Central Valley, uncommon elsewhere in its range; to 4,900 ft (1,500 m). Range: S Mexico to CR.

Red-capped Manakin
Ceratopipra mentalis

male

female

4 in (10 cm). Male has vivid red-orange head and white iris; when displaying, exposes bright yellow thighs. Female is uniformly drab-olive; note short tail and pinkish basal half of lower mandible. Favors mature wet forest, but also occurs in second growth and sometimes gardens; feeds at lower and middle levels. Produces a looping, drawn-out *sp, sp, spweeeeee*, followed by a sharp *spip!*. Common in wet lowlands and uncommon in foothills; to 3,600 ft (1,100 m). Range: SE Mexico to NW Ecuador.

White-collared Manakin
Manacus candei

male

female

4 in (10 cm). The boldly contrasting pattern and bright orange legs distinguish the male. Female is olive above, but has yellow belly and orange legs. Stays in lower levels of second growth and forest edges. Displaying males produce loud snapping and popping noises with their wings, accompanied by short, whistled vocalizations. Common in Caribbean lowlands and foothills; to 3,000 ft (900 m). Range: S Mexico to W Panama.

male

female

Orange-collared Manakin
Manacus aurantiacus

4 in (10 cm). The orange collar sets apart the diminutive male; female and juvenile of this species are virtually identical to those of White-collared Manakin (no range overlap). Female lacks protruding central tail feathers of female Long-tailed Manakin (p. 175). Inhabits lower levels of mature wet forest, second growth, and forest edge. Calls with a slightly rising *chewwe*; in addition to wing snapping, displaying males also make a buzzing noise. Fairly common in southern Pacific lowlands and foothills; to 3,600 ft (1,100 m). Range: CR and W Panama.

Cotingas (COTINGIDAE). No other family of birds in the world contains such a mixed assortment of species, ranging in size from one of the smallest to the largest neotropical passerines. Exhibiting a variety of morphological features, behavior, and nest structures, perhaps the sole shared feature among these birds is their mystery and intrigue. Many species have very limited ranges. Most are predominantly frugivorous, and the males of many species are thus able to devote a lot of time to sexual display. The nests of nearly half of the family remain undescribed and, in general, there is still much to be learned about this fascinating group of birds. World: 66, CR: 8

Bare-necked Umbrellabird

Cephalopterus glabricollis

male

Male 17 in (43 cm); female 14 in (36 cm). The male is quite unique, with his "mod hair-do" and red throat sac (though this is only inflated during display). Female and juvenile birds can be indentified by black coloration, large size, and a suggestion of a floppy crest over the forehead. Inhabits lower and middle levels of mature wet forest. Generally silent, though wings produce a deep rustling in flight; displaying males emit a deep, hollow *huuU!* Uncommon on Caribbean slope from Miravalles Volcano south. Breeds from March to June at middle elevations, from 2,600 to 6,600 ft (800 to 2,000 m); then migrates down to foothills and contiguous lowlands, from 160 to 1,600 ft (50 to 500 m). Range: CR and W Panama.

Turquoise Cotinga
Cotinga ridgwayi

male

female

7 in (18 cm). The stunning male is distinctive in his range. (Very similar Lovely Cotinga *C. amabilis* [not illustrated] is a rare inhabitant of Caribbean foothill forests.) The gray-brown female looks more speckled than any dove or thrush that she might be mistaken for. Perches and feeds in canopy of mature wet forest, as well as in partially cleared areas. Male in flight produces a soft, high-pitched tinkling sound. Uncommon in southern Pacific; to 5,900 ft (1,800 m). Range: CR and W Panama.

Yellow-billed Cotinga
Carpodectes antoniae

male

8 in (20 cm). The yellow bill is the definitive field mark on both the nearly pure-white male and the more grayish female. (Very similar Snowy Cotinga *C. nitidus* [not illustrated] is uncommon in Caribbean lowlands and has grayish bill.) Actively forages for fruit in canopy of mature wet forest and in tall trees of adjacent open areas; also found in mangroves. Generally silent. Uncommon in southern Pacific lowlands and rare in adjacent foothills; to 2,600 ft (800 m). Range: CR and W Panama.

Three-wattled Bellbird
Procnias tricarunculatus

male

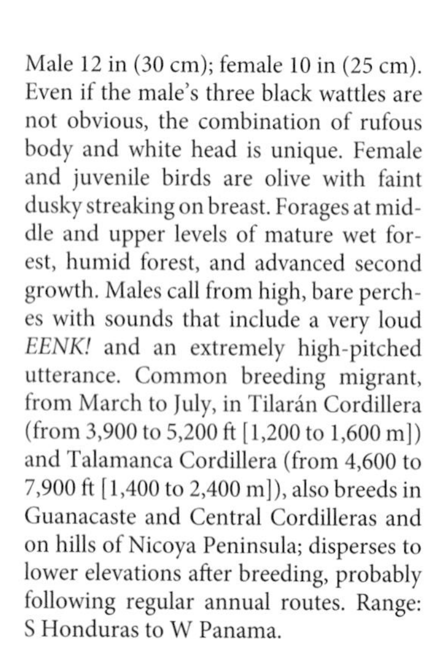

Male 12 in (30 cm); female 10 in (25 cm). Even if the male's three black wattles are not obvious, the combination of rufous body and white head is unique. Female and juvenile birds are olive with faint dusky streaking on breast. Forages at middle and upper levels of mature wet forest, humid forest, and advanced second growth. Males call from high, bare perches with sounds that include a very loud *EENK!* and an extremely high-pitched utterance. Common breeding migrant, from March to July, in Tilarán Cordillera (from 3,900 to 5,200 ft [1,200 to 1,600 m]) and Talamanca Cordillera (from 4,600 to 7,900 ft [1,400 to 2,400 m]), also breeds in Guanacaste and Central Cordilleras and on hills of Nicoya Peninsula; disperses to lower elevations after breeding, probably following regular annual routes. Range: S Honduras to W Panama.

female

Jays (CORVIDAE). This essentially cosmopolitan family is typified as much by its "attitude" as by any observable morphological feature. Corvids are renowned for being among the world's most intelligent animals, and communication plays an important role in their social structure. Few species produce very musical sounds, but what they may lack in quality, they more than make up for with quantity of vocalizations. They are generally curious creatures and are quick to mob predators. Nests are usually open cups placed on top of layers of larger sticks. Both of the species treated here are cooperative nesters, with additional birds assisting the nesting pair with feeding and defense. World: 125, CR: 5

White-throated Magpie-Jay
Calocitta formosa

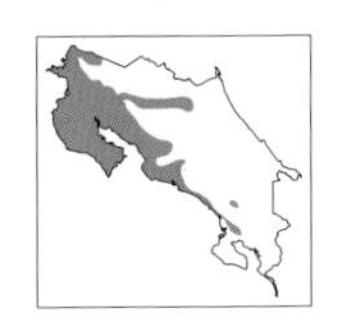

19 in (48 cm). Very attractive with its blue upperparts, long tail, and curled crest feathers. Noisy groups travel through both forested and more open areas, including gardens. Varied vocalizations range from raspy scolding notes to sweet oriole-like chirping. Common throughout the northern Pacific; to 3,900 ft (1,200 m). Seems to be extending its range south along the Pacific coast, with sightings south of Dominical; also spreading in northern central region, with reports from the area between the Arenal Volcano and Guápiles; has entered western Central Valley (to Santa Ana and Belén). Range: S Mexico to CR.

adult

Brown Jay
Psilorhinus morio

juvenile

15 in (38 cm). A fairly large, dark bird with white on belly and on tips of underside of tail feathers. Juvenile has orange bill and eye ring. Boisterous parties roam through trees, avoiding mature forest habitats. Gives a rather shrill and agitated *piyah, piyah*; makes an audible popping noise with its special throat sac. Common from foothills to middle elevations, uncommon in lowlands, rare or absent on much of southern Pacific slope; to 7,500 ft (2,300 m). Range: SE Texas to W Panama.

Thrushes, Allies (TURDIDAE). Though many of the Old World species formerly considered thrushes (e.g., the Common Nightingale *Luscinia megarhynchos*) have now been placed in the Muscicapidae (Old World Flycatchers), this is still a large, cosmopolitan family that contains some of the most familiar birds in the world (e.g., the American Robin *Turdus migratorius*). A generalist lifestyle—combining invertebrates and fruit in the diet, along with both terrestrial and arboreal behavior—has led to the successful colonization of most of the world's ecosystems, apart from wetlands. While not all species are gifted singers, those that are have inspired prose and poetry in our species, and contributed to the family's familiarity. Generally monogamous, most thrushes construct open-cup nests in vegetation. Young birds typically show spotted plumage. World: 162, CR: 15

Black-faced Solitaire
Myadestes melanops

7 in (18 cm). No other mostly gray bird has orange bill and legs. Dwells in lower and middle levels of montane wet forests. Most often seen while feeding on berries; extremely difficult to locate when it is singing. Its unhurried, ethereal song is a characteristic feature of montane wet forests, combining pure flute-like tones with a shrill note that sounds like a rusty gate hinge. (If flute-like tones are heard without shrill note, could suggest song of Slaty-backed Nightingale-Thrush [p. 184].) Common at middle and upper elevations, from 3,300 to 9,200 ft (1,000 to 2,800 m); descends to 1,300 ft (400 m) on Caribbean slope in second half of year. Range: CR and W Panama.

Black-billed Nightingale-Thrush
Catharus gracilirostris

6 in (15cm). Sporting the drab gray and brown tones so characteristic of the local representatives of the family, this is the only nightingale-thrush with an entirely dark bill. Inhabits understory of oak forests and adjacent gardens and clearings, also in paramo; often very confiding. Its song, though somewhat muddled, retains the flute-like essence of the genus. Common in highlands, from 7,200 ft (2,200 m) to timberline; fairly common above timberline. Range: CR and W Panama.

Orange-billed Nightingale-Thrush
Catharus aurantiirostris

6 in (15 cm). The only nightingale-thrush that combines an entirely orange bill with russet-brown back and wings. (Southern CR race has distinctly gray crown.) Similar to Ruddy-capped Nightingale-Thrush (p. 184), but has eye ring. Forages at low levels in tangled second growth, shady gardens, and coffee plantations. Song is varied, faster, and less musical than that of other nightingale-thrushes. Fairly common at middle elevations on the Pacific slopes of the Tilarán, Central, and Talamanca Cordilleras, from 1,600 to 7,200 ft (500 to 2,200 m); also on higher hills of Nicoya Peninsula. Range: Mexico to NE Venezuela.

Slaty-backed Nightingale-Thrush
Catharus fuscater

7 in (18 cm). Recalls Sooty Thrush (p. 186) of highlands (no range overlap), but note whitish belly Forages in forest understory; at dawn and dusk, often hops along trails. Song recalls that of Black faced Solitaire (p. 182), but is lower pitched and less complex—lacks final shrill note; typical versio slides up, then down the scale: *tlee-dee, teedle-doo*. Common at middle elevations of Caribbean slope from 2,600 to 5,900 ft (800 to 1,800 m); also common at upper elevations on Pacific slope of Tilará Cordillera and in the Dota region. Range: CR to SE Peru.

Ruddy-capped Nightingale-Thrush
Catharus frantzii

6 in (15 cm). The rufous crown and bicolored bill distinguish it from other nightingale-thrushes (compare with Orange-billed Nightingale-Thrush [p. 183]). Forages close to the ground in montane forests; at dawn and dusk, comes out into the open in adjacent gardens and clearings. Sings varied phrases with distinctive quality of nightingale-thrushes. Common in highlands; from 4,900 to 8,200 ft (1,500 to 2,500 m). Range: Mexico to W Panama.

Swainson's Thrush
Catharus ustulatus

7 in (18 cm). The buffy eye ring and diffusely spotted breast differentiate it from other thrushes with grayish-brown backs. More arboreal than other migrant thrushes; occurs in a wide range of habitats. Spring migrants often sing rising, flute-like phrases. A common and widespread passage migrant, from late Sept to mid-Nov and from late March to mid-May; an uncommon winter resident; to 9,200 ft (2,800 m). Range: Breeds in NA, winters south to N Argentina.

Wood Thrush
Hylocichla mustelina

in (20 cm). Boldly spotted underparts and a bright rufous crown and back set it apart. Usually seen n or near the ground in mature wet forest or advanced second growth. In alarm, utters a quick *pik-ik-pik*. A fairly common migrant, from Oct to April; to 5,600 ft (1,700 m). Range: Breeds in NA, vinters south to Panama.

Sooty Thrush
Turdus nigrescens

male

10 in (25 cm). Similar to Slaty-backed Nightingale-Thrush (p. 184), no range overlap, but has entirely dark underparts. Female is browner. Juvenile is heavily spotted. Forages on ground in open areas and at forest edges. Call is a harsh, jay-like *retret-ret*. Common above 7,900 ft (2,400 m). Range: CR and W Panama.

female

juvenile

Clay-colored Thrush
Turdus grayi

9 in (23 cm). Told apart from all other dull-brown birds by the yellowish bill and reddish iris. A true garden-variety bird that hops on the ground in typical robin fashion; routinely flicks its tail on landing. From March through June, tirelessly whistles melodic phrases that are responsible for its status as the national bird of Costa Rica. Widespread and common, to 7,900 ft (2,400 m), though uncommon in northern Pacific. Range: E Mexico to N Colombia.

White-throated Thrush
Turdus assimilis

9 in (23 cm). The streaked throat, bordered below by a white crescent, distinguishes it from other species with yellow-orange bill, eye ring, and legs. Prefers middle levels of mature forests and adjacent second growth. A talented vocalist that sings surprisingly varied phrases ranging from tuneful whistles to thin trills, all typically given in couplets. Fairly common on Pacific slope, uncommon on Caribbean slope; from 2,600 to 5,900 ft (800 to 1,800 m). Some birds move to lower elevations from June to Dec. Range: Mexico to NW Ecuador.

Dippers (CINCLIDAE). The dippers are peculiar among passerines, as the two Old World species and the American Dipper possess the unique ability to forage underwater (the two South American species apparently do not). All five species occur along fast-flowing streams. They feed chiefly on the larvae and nymphs of aquatic insects, but also consume mollusks and small fish. Mostly monogamous, pairs nest over water on boulders, cliffs, and man-made structures. The rounded mass of moss is lined with grass and leaves. Young birds are capable of swimming, even before flying. World: 5, CR: 1

American Dipper
Cinclus mexicanus

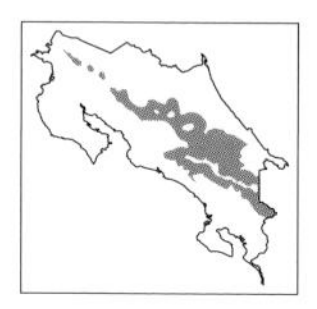

7 in (18 cm). The overall gray coloration, plump body, short tail, and fairly long, pale legs impart a distinctive look. (Compare with Torrent Tyrannulet [p. 155], found in same habitat.) Forages on river rocks for aquatic insects, sometimes disappearing below the water's surface; habitually bobs hindparts; flies low along streams. Call is a sharp, buzzy *dzeet*. Fairly common in middle elevations and highlands of Caribbean slope, from 2,300 to 8,200 ft (700 to 2,500 m); uncommon on Pacific slope of Central and Talamanca Cordilleras, from 5,600 to 8,200 ft (1,700 to 2,500 m). Range: Alaska to W Panama.

Mockingbirds (MIMIDAE). The members of this New World family typically avoid dense forest, though not necessarily dense vegetation. The majority are fairly terrestrial and have correspondingly long legs and tail. Arthropods comprise the bulk of the diet, which is supplemented with fruit. The sexes are alike and form monogamous pairs that vigorously defend a territory. Being quite vocal is part of their pair bonding and defense strategy. Some species are accomplished mimics. Pairs construct an open-cup nest, the female does the incubation, and both parents participate in feeding the young. World: 34, CR: 2

Tropical Mockingbird
Mimus gilvus

10 in (25 cm). The long tail, pointed bill, and pale iris might suggest a female grackle (pp. 236-237), but note the narrow white wing bars and white-tipped tail. Prefers open areas (including fields, soccer fields, and urban gardens). An accomplished singer; despite its name, it rarely imitates other species. However, in CR it has been heard doing renditions of both Common Pauraque (p. 79) and Clay-colored Thrush (p. 187). Since the turn of the century, has become widespread and definitely established in many sites, but is still generally uncommon; to 5,600 ft (1,700 m). Range: S Mexico to E Brazil.

Silky-flycatchers (PTILIOGONATIDAE). This New World family consists of two species endemic to the highlands of Costa Rica and western Panama, another confined to the highlands of Mexico and western Guatemala, and a fourth that inhabits deserts in the southwestern US and in Mexico. The common name is misleading, as they are not closely related to flycatchers, and consume far more fruit than insects—the latter mostly during breeding season. Females are duller than the respective males in each species. Pairs construct an open-cup nest and both members participate in nesting duties. World: 4, CR: 2

Black-and-yellow Silky-flycatcher
Phainoptila melanoxantha

male

female

8 in (20 cm). Both sexes told apart from any of the thrushes (pp. 182-187) by yellow flanks (male also has yellow rump). Birds of the Guanacaste and Tilarán Cordilleras have more extensive yellow on underparts. Rather sedentary; feeds on berries at middle levels and in canopy of montane forest and forest edge. Gives a high, thin *tsip*. Fairly common in highlands; above 3,900 ft (1,200 m) in Guanacaste and Tilarán Cordilleras, and from 5,900 ft (1,800 m) to timberline in Central and Talamanca Cordilleras. Range: CR and W Panama.

male

Long-tailed Silky-flycatcher

Ptiliogonys caudatus

female

9 in (23 cm). The prominent crest, sleek proportions, and soft gray-and-yellow coloration (female is more olive) set it apart. Found in forests, forest edges, and gardens; typically perches upright at tips of branches located near the top of tall trees, but can be seen as low as eye level when it feeds in fruiting shrubs. Makes a wooden and staccato (but sweet) chittering. Fairly common in Central and Talamanca Cordilleras; from 5,200 ft (1,600 m) to timberline. Range: CR and W Panama.

Gnatwrens, Gnatcatchers (POLIOPTILIDAE). A New World family of small insectivores characterized by a long, thin bill and—in most species—a long, narrow tail. They occupy a range of habitats, from desert scrub to tropical rain forest, but all forage actively for small arthropods gleaned from vegetation. Despite being considered somewhat closely related to wrens, few utter sounds that could be termed musical. Pairs construct a small cup-shaped nest, with both parents sharing in incubation, brooding, and feeding. World: 16, CR: 4

Long-billed Gnatwren

Ramphocaenus melanurus

5 in (13 cm). The long, slender bill and tail, combined with a pale, tawny face are distinctive. Pokes about in thickets and tangled vines at forest edges and openings, and in adjacent second growth. Frequently sings a staccato trill, which can vary in quality from wooden to dulcet lasting about two seconds and often rising slightly at the end. Common in humid areas; to 3,900 ft (1,200 m); restricted to evergreen patches in northern Pacific. Range: SE Mexico to N Bolivia.

Tropical Gnatcatcher

Polioptila plumbea

male

4 in (10 cm). The combination of gray upperparts and white underparts, together with a slender appearance, is indicative of a gnatcatcher. Both sexes show a broad white superciliary, whereas the very similar White-lored Gnatcatcher *P. albiloris* (not illustrated), which is confined to northern Pacific dry forests, has a narrow white superciliary. Pairs forage actively at all levels in second growth, forest edges, and gardens; often holds tail cocked up. Gives a nasal *myew*; song is a descending series of fast, high chirps. Common in wet lowlands and middle elevations, to 3,900 ft (1,200 m); uncommon in northern Pacific and western Central Valley. Range: SE Mexico to Peru.

female

Wrens (TROGLODYTIDAE). Essentially a New World family—just one species inhabits Europe and Asia. The largely insectivorous wrens exploit a wide range of habitats, as is indicated by a sampling of their common names: Cactus Wren, Marsh Wren, Timberline Wren, and House Wren. The majority of species, however, dwell in the lower strata of wooded environments, where their generally rather discreet attire serves them well. Their vocalizations, in contrast, are often quite attention grabbing, and even astonishing. Nests are rounded, covered structures, though a few species are cavity nesters that build cup-shaped nests inside a hollow. Most species breed as monogamous pairs, but there are also polygamists and those that are assisted by the young of previous clutches. World: 82, CR: 22

House Wren

Troglodytes aedon

4 in (10 cm). Told apart from other small, drab-brown birds of open habitats by the barring on the wings and tail. Usually around buildings or in weedy pastures. Vocalizes from an exposed perch; utters a spirited, bubbly song that mixes musical trills and chattering (similar to song of Tropical Parula [p. 203]). Common and widespread, to 9,200 ft (2,800 m); uncommon in northern Pacific. Range: S Canada to Tierra del Fuego.

Timberline Wren

Thryorchilus browni

4 in (10 cm). The white markings on the folded primaries are the definitive field mark on this otherwise "generic" wren. (The only other wren at all likely in its limited highland range is the Gray-breasted Wood-Wren [p. 197].) Forages in bamboo and dense shrubbery. The song, neither full-bodied nor exceedingly musical, is a rapid delivery of high warbles and trills. Fairly common in highlands above 8,900 ft (2,700 m). Range: CR and W Panama.

Band-backed Wren
Campylorhynchus zonatus

7 in (18 cm). The bold patterning should eliminate confusion (the Spot-breasted Wren *Pheugopedius maculipectus* [not illustrated] has a plain, brown back and occurs only in the Caño Negro region). Forages on branches and trunks of trees at middle and lower levels of forest edges and gardens; often conspicuous. Utters a squeaky, scratchy series of notes that have a certain lilting rhythm. Fairly common on Caribbean slope; to 3,900 ft (1,200 m). Range: Mexico to NW Ecuador.

Rufous-naped Wren
Campylorhynchus rufinucha

7 in (18 cm). Unique among local wrens, this species exhibits both a bold pattern and bold behavior. Forages at middle and lower levels of dry forest, second growth, and gardens, often around human habitations. Pairs and family members call back and forth with a potpourri of both musical and grating notes. The bulky straw nests are often seen in small, thorny ant-acacia trees. Common in northern Pacific and into the western half of the Central Valley; to 3,300 ft (1,000 m). Range: Mexico to CR.

Stripe-breasted Wren
Cantorchilus thoracicus

5 in (13 cm). The only CR wren with a streaked breast. Pairs forage in lower levels of mature wet forest and forest edges. Delivers a series of clear, even-pitched whistles that recall the vocalization of a pygmy-owl (p. 74). Also quickly repeats a short phrase of dulcet tones, then switches to another phrase. Common in Caribbean lowlands and foothills; to 3,300 ft (1,000 m). Range: E Nicaragua to W Panama.

Plain Wren
Cantorchilus modestus

5 in (13 cm). Distinguished by the combination of buffy flanks and white superciliary, throat, and breast. The duller and grayer birds of the Caribbean lowlands were once considered a separate species: Canebrake Wren *C. zeledoni*. Forages at lower levels of overgrown vegetation in nonforest habitats. The CR common name derives from its typical song: *chinchirigüí*. Common on Pacific slope and in the Central Valley, to 6,600 ft (2,000 m); fairly common in Caribbean lowlands; uncommon in northern Pacific lowlands. Range: S Mexico to W Panama.

Riverside Wren
Cantorchilus semibadius

5 in (13 cm). The only bird in CR that combines finely barred black-and-white underparts with a rich rufous crown and back. Pairs or small groups forage low in dense vegetation inside mature wet forests and at forest edges (despite its name, not confined to riversides); regularly comes into view. Has loud, sweet, fast-paced songs that are not quite as explosive as those of the Bay Wren. Common in southern Pacific; to 3,900 ft (1,200 m). Range: CR and W Panama.

Bay Wren
Cantorchilus nigricapillus

6 in (15 cm). The chestnut body and white throat and facial markings readily distinguish it (like most wrens, however, it can be difficult to see). Almost always found near watercourses, working its way through thick shrubbery. The powerful voice veritably explodes from the underbrush with rapidly repeated phrases. Common in Caribbean lowlands and foothills; to 3,300 ft (1,000 m). Range: SE Nicaragua to Ecuador.

White-breasted Wood-Wren
Henicorhina leucosticta

4 in (10 cm). The white throat and breast are the key field marks on this small, brown-backed wren. Active and noisy, it mostly forages on or near the ground in mature wet forest and adjacent advanced second growth. Produces an often heard chattering, a piercing *cheet* call note, and quite a variety of clear, whistled phrases (similar to song of Gray-breasted Wood-Wren). Common in Caribbean lowlands and foothills, to 3,900 ft (1,200 m); common at middle elevations in southern Pacific, to 5,900 ft (1,800 m); rare in southern Pacific lowlands. Range: E Mexico to central Peru.

Gray-breasted Wood-Wren
Henicorhina leucophrys

4 in (10 cm). No other wren combines black-and-white facial markings and a gray breast. Dwells in understory of wet montane forest and forest edges. Its loud, lengthy, rollicking melody is one of the most commonly heard and attention grabbing sounds in its habitat. Utters a chatter similar to that of the White-breasted Wood-Wren. Abundant from 2,600 ft (800 m) to timberline. Range: Mexico to W Bolivia.

Vireos (VIREONIDAE). With the recent inclusion of ten Asian species (Shrike-Babblers), this is no longer an exclusively New World family. Dull green is a prominent color in most species and, together with their average size, can lead to confusion with New World Warblers (pp. 200-209); but vireos are typically slower moving and have thicker bills than warblers. Most species have a hook-tipped bill. Arthropods taken from the surface of vegetation account for much of the diet, though small fruits are also important, especially outside of the breeding season. The term that perhaps best sums up their vocalizations is *persistent*. Sexes are alike in the western hemisphere, where a pair builds a cup nest that is suspended from its rim in a horizontal fork of a branch. World: 62, CR: 16

Yellow-throated Vireo

Vireo flavifrons

5 in (13 cm). The combination of white wing bars and yellow spectacles, throat, and breast is definitive. Single individuals can be found in a variety of habitats, but typically in gardens and second growth, and at forest edges; often accompanies mixed species flocks. Makes a descending series of harsh, fast notes: *cheh-cheh-cheh-cheh-chehcheh*. A common and widespread migrant, from late Sept to late April; to 5,900 ft (1,800 m). Range: Breeds in E NA, winters south to Venezuela.

Yellow-winged Vireo

Vireo carmioli

4 in (10 cm). Only the Yellow Tyrannulet *Capsiempis flaveola* (not illustrated), a lowland flycatcher shares an olive back and yellow superciliary, wing bars. and underparts. Forages at all levels in mature forests, forest edges, and gardens; often accompanies mixed species flocks. Sings a leisurely series of two- and three-note phrases, with distinct pauses between each phrase. Common in highlands of Central and Talamanca Cordilleras; from 4,900 ft (1,500 m) to timberline. Range: CR and W Panama.

Yellow-green Vireo
Vireo flavoviridis

6 in (15 cm). Closely resembles Red-eyed Vireo *V. olivaceus* (not illustrated), a common passage migrant from mid-Aug to late Nov and from late March to late May, but the latter has white superciliary strongly outlined by black above and below, and shows almost no yellow on underparts. Forages at middle and upper levels of second growth and gardens. Sings short simple phrases repetitiously, often while hidden in vegetation. Common breeding migrant on Pacific slope, rare on Caribbean slope; from lowlands to about 4,900 ft (1,500 m). Migrates to South America after breeding (most gone by late Oct), begins returning in late Jan. Range: Breeds from S Texas to Panama, winters south to N Bolivia.

Lesser Greenlet
Hylophilus decurtatus

4 in (10 cm). Brings to mind a Tennessee Warbler (p. 202) in coloration, size, and behavior, but note narrow, white eye ring. Forages actively in middle and upper levels of forests, forest edges, and advanced second growth; usually in small groups and with mixed flocks. Repetitiously and leisurely sings a short, warbled phrase: *chi-uree.* Common and widespread; to 4,900 ft (1,500 m). Range: E Mexico to N Peru.

New World Warblers (PARULIDAE). The great majority of New World warblers are small, active insectivores that forage mostly by gleaning from vegetation. Outside of the breeding season, some species also feed on nectar and fruit. All species breeding in the temperate zone are migratory, and in many cases the males are more strikingly colored than their mates; some species molt into a drabber plumage after breeding. Sexual dimorphism is slight among the non-migratory tropical species and there are no nonbreeding plumages. Most species build open-cup nests, though some construct dome-shaped nests, and a few nest in cavities. The female constructs the nest and incubates the eggs, while the male helps in feeding and defense of the nest. World: 114, CR: 53

Northern Waterthrush
Parkesia noveboracensis

6 in (15 cm). A superciliary that tapers slightly at the rear, and flecking on the throat, are the two field marks that distinguish it from the otherwise similar Louisiana Waterthrush *P. motacilla* (not illustrated), which is usually found along fast-flowing streams at higher elevations. Terrestrial, prefers edges of ponds and slow-moving streams, also trails; habitually teeters as it walks or stands. Utters a metallic *tsink*. Common migrant, from mid-Aug to mid-May, mostly in lowlands; to 4,900 ft (1,500 m). Range: Breeds in N NA, winters south to N SA.

Black-and-white Warbler
Mniotilta varia

female

5 in (13 cm). No other bird in CR is black-and-white striped above and below; also note habit of creeping on tree trunks and branches. Forages at middle and lower levels of forest, forest edges, and gardens; often with mixed species flocks. Common migrant, from mid-Aug to mid-April, most numerous at middle elevations and in highlands; to 8,200 ft (2,500 m). Range: Breeds in E NA, winters south to Peru.

Prothonotary Warbler
Protonotaria citrea

male

female

5 in (13 cm). Distinguished from other bright yellow birds by combination of white undertail coverts and gray wings and rump. Females show yellow that is slightly duller. Forages in lower levels of vegetation, usually near water (especially mangroves). Fairly common migrant, from mid-Aug to late March, mostly in lowlands; to 4,900 ft (1,500 m). Range: Breeds in E US, winters south to W Ecuador.

Flame-throated Warbler
Oreothlypis gutturalis

5 in (13 cm). The orange throat (slightly duller in females and juveniles) and black back are a unique combination on an otherwise gray-and-white bird. Pairs or small groups forage at middle and upper levels of forests, forest edges, and adjacent gardens; often accompanies mixed species flocks. Gives a high, drawn-out, buzzy, arcing note. Fairly common in highlands of Central and Talamanca Cordilleras; from 5,200 ft (1,600 m) to timberline. Range: CR and W Panama.

Tennessee Warbler
Oreothlypis peregrina

breeding

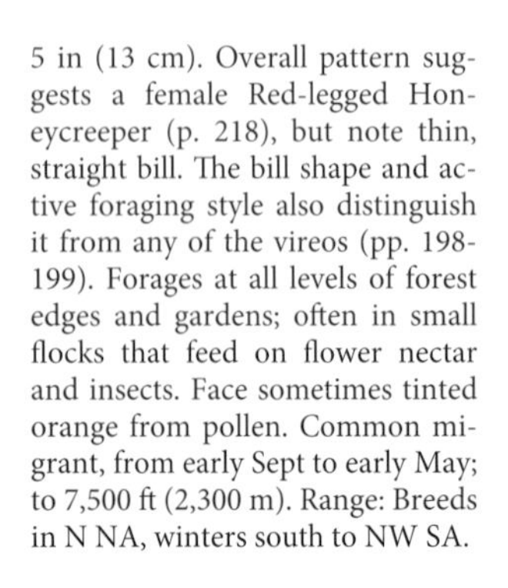

5 in (13 cm). Overall pattern suggests a female Red-legged Honeycreeper (p. 218), but note thin, straight bill. The bill shape and active foraging style also distinguish it from any of the vireos (pp. 198-199). Forages at all levels of forest edges and gardens; often in small flocks that feed on flower nectar and insects. Face sometimes tinted orange from pollen. Common migrant, from early Sept to early May; to 7,500 ft (2,300 m). Range: Breeds in N NA, winters south to NW SA.

nonbreeding

Gray-crowned Yellowthroat
Geothlypis poliocephala

male

5 in (13 cm). The small black mask, gray crown, and yellow throat readily differentiate the male. (Males of the other three species of yellowthroat found in CR all have more extensive black masks that extend down to the cheeks and sides of throat.) The female and juvenile can be identified by the relatively thick, bicolored bill and long tail. Forages low in brushy fields. From atop shrubbery, male sings sweet, wrenlike phrases. Common and widespread; to 4,900 ft (1,500 m). Range: S Texas to W Panama.

Tropical Parula
Setophaga pitiayumi

4 in (10 cm). The bluish upperparts and bright yellow underparts (with an orangish wash across the breast) are distinctive. Forages at middle and upper levels of forest edges, second growth, and gardens; often picks at petiole bases of *Cecropia* leaves. Sings a series of high, fast notes and trills that recall song of House Wren (p. 193). Fairly common on Caribbean slope, fairly uncommon on Pacific slope; from 2,000 to 5,900 ft (600 to 1,800 m); has been reported at sea level near Carara NP. Range: S Texas to N Argentina.

[There are two populations of Yellow Warbler in mainland CR: a resident population, often referred to as Mangrove Warbler, and NA migrants. For the sake of clarity, each is given a separate account, though they are currently considered the same species.]

Yellow Warbler (Mangrove)

Setophaga petechia

male

female

5 in (13 cm). Adult male is easily distinguished by its entirely chestnut head; female has chestnut center of crown; juvenile is rather drab. Forages at all levels in mangrove swamps; occasionally in adjacent wooded habitats. Sings a quick, sweet *whe-wheche-wechu-wee*. Common inhabitant of Pacific coast mangrove swamps; rare in Caribbean mangroves north of Moín. Range: Mexico to N Peru and N Venezuela, including Caribbean islands.

Yellow Warbler (Northern)

Setophaga petechia

male

female

5 in (13 cm). The rufous-streaked breast readily identifies the bright male. Female lacks streaking and is slightly duller (juvenile females are quite pale); flashes yellow in the tail. Active and noisy; forages at lower and middle levels of gardens. Emits a near constant, sharp *chip*. Common migrant, from mid-Aug to early May; to 4,900 ft (1,500 m). Range: Breeds in NA and winters south to Ecuador.

Chestnut-sided Warbler
Setophaga pensylvanica

breeding male

nonbreeding

5 in (13 cm). The most commonly encountered migrant warbler in wet forest habitats. Distinctive breeding plumage is only observed from about late March onward. In nonbreeding plumage, note the olive upperparts, pale underparts, white eye ring, and pale yellow wing bars. Typically holds tail cocked up and wings drooped down. Forages at all levels of forests, forest edges, second growth, and gardens; often with mixed flocks. Gives a sharp *chirp*. Common migrant (though scarce in northern Pacific), from early Sept to mid-May; to 7,200 ft (2,200 m). Range: Breeds in E NA, winters south to Panama.

Rufous-capped Warbler
Basileuterus rufifrons

5 in (13 cm). The facial pattern is diagnostic in telling it apart from other warblers with bright yellow underparts. Forages in lower levels of woodlands, second growth, forest edges, and shaded coffee plantations; typically holds tail cocked upwards. Sings a spritely jumble of accelerating thin, high notes. Common in northern Pacific and in Central Valley, east to Turrialba; to 6,600 ft (2,000 m). Also common in intermontane valleys of southern Pacific; from 2,000 to 5,200 ft (600 to 1,600 m). Range: Mexico to W Venezuela.

Black-cheeked Warbler
Basileuterus melanogenys

5 in (13 cm). Overall size, coloration, and habits could cause confusion with the sympatric Sooty-capped Chlorospingus (p. 230), but note the rufous crown patch. Forages actively at lower levels of highland forest and at forest edge; usually two or more birds are present (often with mixed flocks). Song similar to song of Rufous-capped Warbler (p. 205), but thinner. Fairly common in highlands of Central and Talamanca Cordilleras; from 5,200 ft (1,600 m) to above timberline. Range: CR and W Panama.

Three-striped Warbler
Basileuterus tristriatus

5 in (13 cm). The combination of head stripes and dark ear coverts differentiates it from other small olive-colored birds. Small groups forage actively in lower levels of mature wet forest and advanced second growth, where, together with Common Chlorospingus (p. 229), they are often the core species within mixed species flocks. Produces a rapid, high-pitched twittering. Common at middle elevations, from Tilarán Cordillera south; from 3,300 to 7,200 ft (1,000 to 2,200 m). Range: CR to Bolivia.

Buff-rumped Warbler
Myiothlypis fulvicauda

5 in (13 cm). The pale superciliary distinguishes it from the similarly-patterned Sulphur-rumped Flycatcher (p. 159). Hops and flits along edges of forest streams and on trails and lawns; often attends army-ant swarms; swishes tail from side to side. Sings an accelerating series of loud, piercing notes that can easily be heard above the sound of rushing water. Common in wet lowlands and foothills; to 3,300 ft (1,000 m) on Caribbean slope, and to 4,900 ft (1,500 m) in southern Pacific. Range: N Honduras to NW Bolivia.

male

Wilson's Warbler
Cardellina pusilla

female

4 in (10 cm). The male's black cap is diagnostic. The female resembles the female Yellow Warbler (p. 204), but has an olive crown and no yellow in the tail. Forages actively at all levels in forests, forest edges, second growth, and gardens; often with mixed flocks. Common winter resident (from early Sept to early May) and passage migrant, mostly above 2,600 ft (800 m); rare passage migrant at lower elevations. Range: Breeds in N NA, winters south to Panama.

Slate-throated Redstart
Myioborus miniatus

5 in (13 cm). The dark face and throat set it apart from the Collared Redstart; also note white in outer tail feathers. Forages in middle and lower levels of mature forest, second growth, and forest edges; usually in pairs and often with mixed species flocks; frequently makes short sallies and fans its tail. Sings a simple song of six or more even notes that increase slightly in pace and volume before stopping abruptly. Common at middle elevations; from 2,300 to 6,900 ft (700 to 2,100 m). Range: Mexico to Bolivia.

Collared Redstart
Myioborus torquatus

5 in (13 cm). The dark collar separating the bright yellow face and underparts, along with the rufous crown, readily identifies this often confiding little bird. Forages in middle and lower levels of mature forest, second growth, and forest edges; usually in pairs and often with mixed species flocks; frequently makes short sallies and fans its tail. Common in highlands, from Tilarán Cordillera south; from 4,900 ft (1,500 m) to timberline. Range: CR and W Panama.

Wrenthrush
Zeledonia coronata

5 in (13 cm). The bright russet crown almost seems out of place on this otherwise somber plumaged species. Stays on or near the ground in dense undergrowth of montane forests and paramo. Hard to see; presence best revealed by thin, penetrating call (*tsseee!*) or high-pitched song (*tsee-te-dee*). Fairly common in highlands; from 4,900 ft (1,500 m) to above timberline. Range: CR and W Panama.

Tanagers, Seedeaters, Allies (THRAUPIDAE). Throughout this and other guides, the reader is advised to look closely at the bill of the bird in question to help determine to which family it belongs. However, recent DNA analysis has resulted in numerous changes in the composition of the tanager family, making that advice rather useless in this case. The evolution of this family has produced bill shapes ranging from that of honeycreepers to flowerpiercers to seed-finches—each designed to facilitate feeding on certain foods. Nor are there any other morphological or behavioral features that readily define a "tanager." Vocalizations likewise vary widely throughout this neotropical family. Perhaps the one area where there is general similarity is in breeding behavior, with most species forming monogamous pairs. The female builds an open-cup nest and incubates the eggs; the male later helps with feeding the young a diet consisting largely of insects. World: 386, CR: 52

Tawny-crested Tanager
Tachyphonus delatrii

male

6 in (15 cm). The male's bright tawny-orange crest feathers are conspicuous. The nondescript brown female is best identified by the company she keeps—she often travels in flocks of a dozen or more individuals, including distinctive males. Moves through understory and lower canopy of mature wet forest in noisy, active flocks that are often joined by other species. Call notes are sibilant and squeaky. Common in Caribbean foothills, from 1,000 to 2,600 ft (300 to 800 m); uncommon from 2,600 to 3,900 ft (800 to 1,200 m); also uncommon in Caribbean lowlands. Range: E Honduras to W Ecuador.

female

White-lined Tanager
Tachyphonus rufus

male

female

7 in (18 cm). Often travels in pairs; since neither male nor female is particularly distinct, the best key to identification is their mere co-presence. In flight, the male flashes a bold white wing patch (normally obscured when perched). The bill shape distinguishes the rufous female from other similarly colored species; also note pale gray on lower mandible of both sexes. Forages in open areas with brushy habitat. Not very vocal, makes a sweet *chew-wE, chewwE…*, the last part upwardly inflected. Fairly common in Caribbean lowlands and foothills (to 1,400 m [4,600 ft] in eastern Central Valley); recently expanding from Panama into southern Pacific. Range: S Nicaragua to N Argentina.

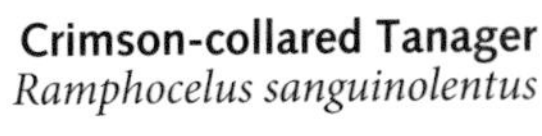

Crimson-collared Tanager
Ramphocelus sanguinolentus

7 in (18 cm). The crimson hood readily distinguishes it from the male Passerini's Tanager (p. 212). Sexes are alike. Individuals or pairs typically forage a few meters above the ground (slightly higher than Passerini's Tanager) in disturbed habitats. Song is a series of slow-paced, sibilant notes. Fairly common throughout Caribbean lowlands and foothills; to 3,900 ft (1,200 m). Range: SE Mexico to W Panama.

Passerini's Tanager
Ramphocelus passerinii

male

6 in (15 cm). On male, the bright scarlet rump (on mostly velvety-black body) is striking, but it is the iris color and bill shape that set it apart from the Scarlet-rumped Cacique (p. 239). The bill shape and color are what identify the female. (No geographic overlap with Cherrie's Tanager.) Numerous birds are generally found together staying low in open areas with ample brush. Calls are varied and frequently given, with a sharp *tsip!* often heard. A common and conspicuous resident of Caribbean lowlands and foothills; to 4,900 ft (1,500 m). Range: S Mexico to W Panama.

female

Cherrie's Tanager
Ramphocelus costaricensis

male

6 in (15 cm). The southern Pacific counterpart of Passerini's Tanager, the males of both species are indistinguishable. The female is told apart by the orange breast and rump. (No geographic overlap with Passerini's Tanager.) Found in disturbed habitat with thickets and shrubs. Common in southern Pacific lowlands and foothills; to 5,900 ft (1,800 m). Range: CR and W Panama.

female

Blue-gray Tanager
Thraupis episcopus

6 in (15 cm). No other CR species is light-blue above and below (brightest on the wings and tail). Its simple song is little more than even-pitched squeaks (similar to that of Palm Tanager). One of the most common birds in gardens throughout the country, to 7,200 ft (2,200 m); uncommon in northern Pacific. Range: SE Mexico to SE Peru.

Palm Tanager
Thraupis palmarum

6 in (15 cm). The two-tone coloration (gray body and black flight feathers) sets it apart. In good light, shows an olive sheen. Song similar to that of Blue-gray Tanager, but more rapid. Common in gardens below about 5,900 ft (1,800 m); uncommon in the Central Valley and rare in northern Pacific. Range: E Nicaragua to S Brazil.

Golden-hooded Tanager
Tangara larvata

5 in (13 cm). The golden-tan hood distinguishes it from other birds with brilliant blue coloration. Juvenile is much duller. Typically travels in pairs or small groups; often with mixed species flocks. Characteristic call is a buzzy twittering. A common species of gardens and forest edges, throughout wet lowlands and into middle elevations; to 5,200 ft (1,600 m). Range: S Mexico to NW Ecuador.

Spangle-cheeked Tanager
Tangara dowii

5 in (13 cm). The pale flecking on the cheeks and nape, together with the speckled breast, set it apart from other species with dark upperparts and pale rufous underparts. Juvenile is much duller. Pairs typically travel with mixed species flocks in montane forests and at forest edges. Fairly common from 4,600 to 9,800 ft (1,400 to 3,000 m). Range: CR to W Panama.

Bay-headed Tanager
Tangara gyrola

5 in (13 cm). The combination of reddish-brown head, bright olive back, and turquoise-blue underparts is distinctive. Juvenile is much duller. Favors middle levels of wet forest habitats, second growth, and gardens. Call is a weak *tsit*. Common on southern Pacific slope, to 4,900 ft (1,500 m); fairly common in foothills and middle elevations of Caribbean slope, from 1,300 to 4,900 ft (400 to 1,500 m), and occasionally in Caribbean lowlands. Range: CR to Bolivia.

Emerald Tanager
Tangara florida

5 in (13 cm). The black ear patch distinguishes this brilliant bird from all other bright-green species. Almost exclusively found with mixed species flocks in wet forest and at forest edge. Uncommon in Caribbean foothills; from 1,300 to 3,300 ft (400 to 1,000 m). Range: CR to NW Ecuador.

Silver-throated Tanager
Tangara icterocephala

5 in (13 cm). The silvery-white throat is normally the last thing one notices on this bright yellow bird with a black-streaked back. Juvenile is much duller. Often in flocks, with or without other species, at middle levels of forest and forest edges. Frequently utters a distinctive buzzy *tseet*. Common in wet foothills and middle elevations, from 1,300 to 6,600 ft (400 to 2,000 m); less common in adjacent lowlands and on the Osa Peninsula. Range: CR to NW Peru.

Scarlet-thighed Dacnis
Dacnis venusta

male

5 in (13 cm). Seeing the scarlet thighs (not visible in photo) is hardly necessary to identify the stunning male. Female can be recognized by the combination of pale-blue head, red eye, grayish breast, and buffy belly. Feeds in fruiting trees at forest edges and in gardens. Fairly common in foothills and middle elevations on Caribbean slope (descends to the base of mountains during second half of year). Also fairly common on Pacific slope of Talamanca Cordillera (from 1,300 to 4,900 ft [400 to 1,500 m]) and Tilarán Cordillera (from 4,300 to 4,900 ft [1,300 to 1,500 m]); uncommon on the Osa Peninsula. Range: CR to SW Ecuador.

female

Shining Honeycreeper
Cyanerpes lucidus

male

female

4 in (10 cm). Bright yellow legs render the male unmistakeable. The blue malar stripe and streaking on the breast of the greenish female is diagnostic. Feeds in forest canopy, at forest edge, and in gardens. Fairly common in wet lowlands and foothills; to 3,900 ft (1,200 m). Range: S Mexico to NW Colombia.

Red-legged Honeycreeper

Cyanerpes cyaneus

male

In all plumages, male flashes yellow underwing.

4 in (10 cm). Male is easily told apart from the male Shining Honeycreeper (p. 217) by combination of turquoise crown and red legs. The olive female resembles a Tennessee Warbler (p. 202), but note longer, and slightly curved, bill. After breeding, male molts into a plumage that resembles female, but with black wings. Found in forests, second growth, forest edges, and gardens; typically in small groups. Common on Pacific slope and in northern central Caribbean lowlands and foothills, to 3,900 ft (1,200 m); uncommon in southern Caribbean lowlands. Range: S Mexico to central Bolivia.

nonbreeding male

female

Green Honeycreeper
Chlorophanes spiza

male

female

5 in (13 cm). Turquoise-green male is distinguished by black half-hood and mostly yellow bill. The lime-green female has yellow lower mandible. Found at forest openings, forest edges, second growth, and gardens; often with mixed species flocks. Fairly common in wet lowlands and middle elevations; to 4,900 ft (1,500 m). Range: S Mexico to SE Brazil.

Slaty Flowerpiercer
Diglossa plumbea

male

female

4 in (10 cm). Both the dark-gray male and the olive-gray female are readily differentiated from all other similarly colored birds by the unique bill shape. Uses bill to pierce the base of flowers and hence "rob" the nectar. Forages at forest edge and in gardens. Song is a high, fast twitter. Common at upper elevations, from 3,900 ft (1,200 m) to above timberline. Range: CR and W Panama.

Blue-black Grassquit
Volatinia jacarina

male

4 in (10 cm). Male resembles several male seedeaters, but has a thinner, pointed bill and, in good light, shows distinct bluish sheen. Bill shape and faintly streaked breast distinguish the dull-brown female from other seedeater-type birds. Inhabits brushy fields and areas of tall grass. Male sings a buzzy *tzeeer* that rises and falls—often given as the bird jumps or flies a foot or so into the air. Common in lowlands and foothills; uncommon at middle elevations; to 5,600 ft (1,700 m). Range: N Mexico to N Argentina.

female

Thick-billed Seed-Finch
Sporophila funereus

male

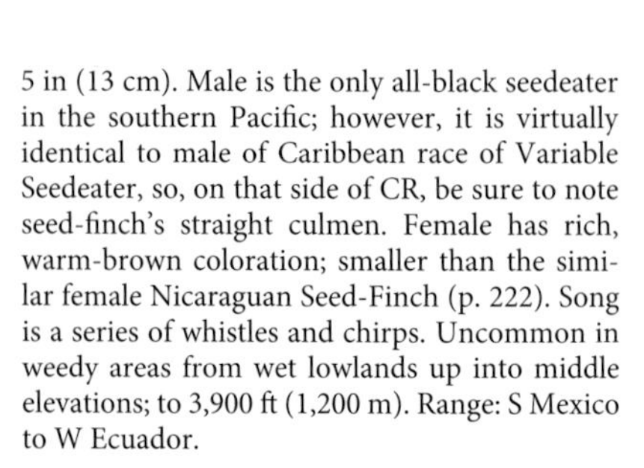

5 in (13 cm). Male is the only all-black seedeater in the southern Pacific; however, it is virtually identical to male of Caribbean race of Variable Seedeater, so, on that side of CR, be sure to note seed-finch's straight culmen. Female has rich, warm-brown coloration; smaller than the similar female Nicaraguan Seed-Finch (p. 222). Song is a series of whistles and chirps. Uncommon in weedy areas from wet lowlands up into middle elevations; to 3,900 ft (1,200 m). Range: S Mexico to W Ecuador.

female

Variable Seedeater
Sporophila corvina

male Pacific race

male Caribbean race

female

4 in (10 cm). Two distinct races occur in CR, with the males looking quite different. Dull-brown female looks the same in both races. Male Pacific race: Could be confused with male White-collared Seedeater *S. torqueola* (not illustrated), but has black throat and lacks wing bars. Male Caribbean race: Virtually identical to male Thick-billed Seed-Finch, except for bill shape (note curved culmen). Inhabits open areas and gardens. Feeds on grass seeds and berries; often can be found surprisingly high in small trees. Common throughout the Caribbean slope, up to 4,900 ft (1,500 m), extending up Reventazón watershed to Cartago. Uncommon in northern Pacific lowlands, increasingly common in southern Pacific, where it can occur up to 4,900 ft (1,500 m). Range: E Mexico to N Peru.

Nicaraguan Seed-Finch
Sporophila nuttingi

male

female

6 in (15 cm). The huge, pink bill renders the male unmistakeable. The rich-brown female has a dark bill; she is bigger—and has a larger bill—than the similarly colored female Thick-billed Seed-Finch (p. 220). Feeds on grass seeds in wet fields that typically have a wooded edge. Sings a medley of squeaky whistles and chirps. Uncommon in Caribbean lowlands. Range: Nicaragua to W Panama.

Bananaquit
Coereba flaveola

4 in (10 cm). The overall color pattern could suggest a kiskadee-type flycatcher (p. 166), but note the thin, decurved bill and small size. Actively works its way through shrubbery and small trees in gardens and at forest edges, feeding on flower nectar, fruits, and insects; comes to hummingbird feeders. Sings a high, scratchy twitter that could be mistaken for a hummingbird. Common from wet lowlands into middle elevations; to 5,200 ft (1,600 m). Range: E Mexico to NE Argentina.

Yellow-faced Grassquit
Tiaris olivaceus

male, dorsal view

male, ventral view

female

4 in (10 cm). Male is unique in having a yellow-orange X on face, bordered by black. Female is drab-olive, but face shows hint of a pale X pattern. Favors fields, roadsides, and gardens. Song is a thin, fast twittering. Common, mostly at middle elevations, from 2,000 to 7,200 ft (600 to 2,200 m); occasionally occurs in lowlands. Range: E Mexico to W Venezuela.

Grayish Saltator
Saltator coerulescens

8 in (20 cm). The white throat, superciliary, and arc below eye, along with pale rufous vent, distinguish it from any other gray bird. Found in gardens and brushy areas with scattered trees. Song, which varies with region, is a pleasant whistled phrase that ends on a rising note. Fairly common in the Central Valley (to 5,900 ft [1,800 m]) and in sections of Caribbean lowlands; uncommon on Pacific slope of Tilarán Cordillera; rare around the Gulf of Nicoya. Range: Mexico to N Argentina.

Buff-throated Saltator
Saltator maximus

8 in (20 cm). The full black bib surrounding the pale-peach throat is distinctive. Also note olive crown. Found in gardens, brushy areas with scattered trees, and at forest edges. Calls with a shrill *tseent*. Sings a melodic series of three notes, each slightly lower pitched: *cheerilee, cheerilu, tsu-tsu*. Common in wet lowlands and into middle elevations, to 5,900 ft (1,800 m); uncommon to rare in northern Pacific and in western Central Valley. Range: E Mexico to Paraguay.

Finches, New World Sparrows, Allies (EMBERIZIDAE). This fairly cosmopolitan family (absent from Australasia) is best represented in the New World, where nearly 75% of its species are found. While the North American sparrows may typify the classic "little brown jobs," many tropical members of the family are olive- or black-backed. Most spend much of their time on or near the ground and have conical bills for cracking open seeds; they also supplement their diet with arthropods. As far as is known, breeding behavior is similar to that of the closely related tanagers (pp. 210-224). Indeed, nearly half of the birds formerly placed in this family are now placed in Thraupidae. World: 172, CR: 25

Yellow-thighed Finch
Pselliophorus tibialis

7 in (18 cm). The dramatic yellow thighs are diagnostic. Juveniles that lack yellow thighs could be puzzlers, but fortunately they are almost always with adults. Pairs or small groups actively forage in mature forest, second growth, forest edges, and gardens; they usually are quite low, but sometimes venture high up into trees. Common from Central Cordillera south, from 4,600 ft (1,400 m) to timberline; less common in higher parts of Tilarán Cordillera. Range: CR and W Panama.

Large-footed Finch
Pezopetes capitalis

8 in (20 cm). The black face and crown stripes on a gray head distinguish this otherwise chunky, olive bird. Often scratches in leaf litter with both feet; occurs in undergrowth of highland forests, second growth, and in paramo. Song is a languid medley of sweet whistles and chirps, rather wren-like in quality. Fairly common from Central Cordillera south; from 6,600 ft (2,000 m) to above timberline. Range: CR and W Panama.

Chestnut-capped Brush-Finch
Arremon brunneinucha

7 in (18 cm). In the dim light of its usual surroundings, the pure-white throat is often more noticeable than the chestnut crown. Inhabits understory of wet forest, advanced second growth, and forest edges. Typically a skulker, but in early morning and late afternoon, sometimes comes out from cover. Common at middle elevations; from 3,000 to 8,200 ft (900 to 2,500 m). Range: E Mexico to SW Ecuador.

Orange-billed Sparrow
Arremon aurantiirostris

adult

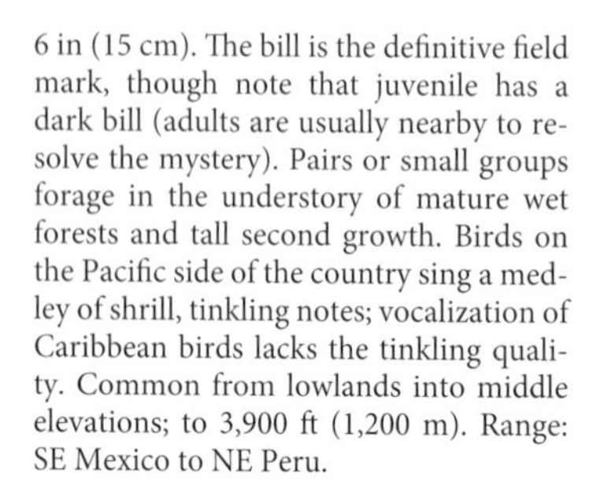

6 in (15 cm). The bill is the definitive field mark, though note that juvenile has a dark bill (adults are usually nearby to resolve the mystery). Pairs or small groups forage in the understory of mature wet forests and tall second growth. Birds on the Pacific side of the country sing a medley of shrill, tinkling notes; vocalization of Caribbean birds lacks the tinkling quality. Common from lowlands into middle elevations; to 3,900 ft (1,200 m). Range: SE Mexico to NE Peru.

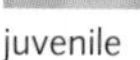

juvenile

Black-striped Sparrow
Arremonops conirostris

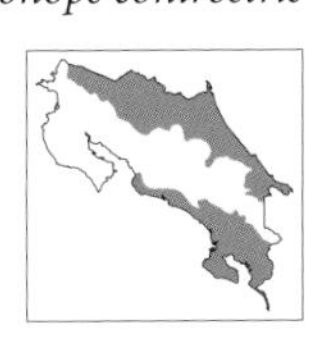

6 in (15 cm). The black stripes on a gray head are diagnostic. The very similar Olive Sparrow *A. rufivirgatus* (not illustrated) of dry forest habitats has dark-brown stripes on a beige head. Favors tangled undergrowth at forest edges, overgrown open areas, and shaded gardens. The attention grabbing song begins slowly with mellow, sweet *chwee* notes that gradually accelerate to a rapid pace toward the end; the performance lasts more than ten seconds. Common in wet lowlands and foothills; to 4,900 ft (1,500 m). Range: E Honduras to N Brazil.

White-eared Ground-Sparrow
Melozone leucotis

7 in (18 cm). The white facial markings and yellow on sides of neck distinguish it from all other olive-backed birds. Found in undergrowth of humid woodlands, heavily shaded gardens, coffee plantations, and along ravines. Song begins slowly, then quickly accelerates into a profusion of loud, high notes. Fairly common from Pacific slope of Tilarán Cordillera to eastern Central Valley; from 2,600 to 6,600 ft (800 to 2,000 m). Range: S Mexico to CR.

Stripe-headed Sparrow
Peucaea ruficauda

7 in (18 cm). The bold head stripes readily differentiate it from all other sparrows with streaked-brown backs. Small groups forage on ground in brushy areas; often seen along fence rows. Song is a squeaky sputtering. Common in northern Pacific lowlands, uncommon in western Central Valley; to 3,300 ft (1,000 m). Sightings near Dominical (2004) suggest it may be spreading south along Pacific coast. Range: NW Mexico to CR.

Rufous-collared Sparrow
Zonotrichia capensis

adult

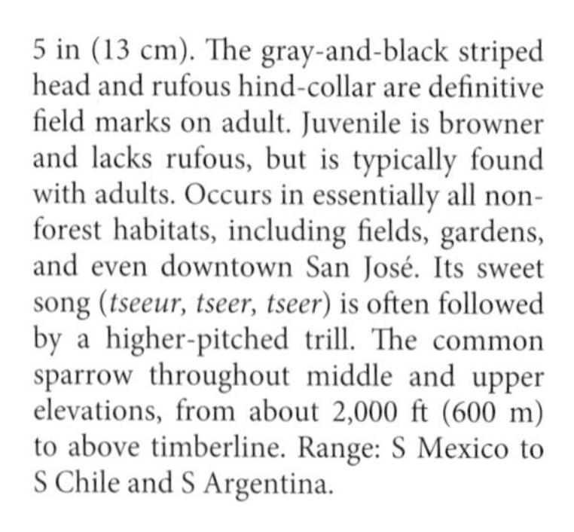

5 in (13 cm). The gray-and-black striped head and rufous hind-collar are definitive field marks on adult. Juvenile is browner and lacks rufous, but is typically found with adults. Occurs in essentially all non-forest habitats, including fields, gardens, and even downtown San José. Its sweet song (*tseeur, tseer, tseer*) is often followed by a higher-pitched trill. The common sparrow throughout middle and upper elevations, from about 2,000 ft (600 m) to above timberline. Range: S Mexico to S Chile and S Argentina.

juvenile

Volcano Junco
Junco vulcani

6 in (15 cm). The orange iris, set off by a small black mask, imparts a fierce look to this highland sparrow; also note pale-orange bill. Typically in pairs that forage on the ground in areas with shrubs and stunted trees, and in paramo. Makes a variety of high, weak utterances. Fairly common above 8,900 ft (2,700 m). Range: CR and W Panama.

Common Chlorospingus
Chlorospingus ophthalmicus

5 in (13 cm). This fairly drab olive-and-gray bird is distinguished by the triangular white spot behind the eye. Actively forages for berries and insects at lower and middle levels of forests and adjacent second growth and gardens; almost always in small groups that are regularly accompanied by other species, especially Three-striped Warblers (p. 206). Noisy and nonmusical, it produces high, weak notes that are often slurred together in a rapid twittering. One of the most common middle-elevation species, from 2,000 to 7,200 ft (600 to 2,200 m). Range: S Mexico to N Argentina.

Sooty-capped Chlorospingus
Chlorospingus pileatus

5 in (13 cm). The jagged white superciliary is the distinctive field mark on this mostly olive-yellow bird. Much more conspicuous than the somewhat similar and sympatric Black-cheeked Warbler (p. 206). Groups actively forage at lower and middle levels of forests and forest edges; often lead mixed flocks. Sings with a thin, squeaky twittering. Common in highlands above 6,600 ft (2,000 m); fairly common on highest peaks of Tilarán Cordillera. Range: CR and W Panama.

Grosbeaks, Buntings, Allies (CARDINALIDAE). Recently, a number of former tanagers were relocated after being shown to be genetically more closely related to the grosbeaks—so, don't be confused by some of the common names that follow. Despite typically having fairly substantial bills that would seem suited to seed-eating, most of these birds also vary their diets with fruit and arthropods. Pleasant songs are a feature of many species in this New World family, as is sexual dimorphism. As might be expected with brightly colored males, females do essentially all of the incubation and brooding in the open-cup nests; however, males often help with feeding the young. World: 48, CR: 20

Flame-colored Tanager
Piranga bidentata

male

7 in (18 cm). The two white wing bars and black-streaked back differentiate both sexes from other species. Prefers forest edges and garden habitats, where individuals or pairs forage at all levels. Call is a resonant, wooden *kriDECK*. Fairly common in highlands on Central and Talamanca Cordilleras, from about 3,900 ft (1,200 m) to timberline. Range: SW Arizona to W Panama.

female

Summer Tanager
Piranga rubra

male

7 in (18 cm). The entirely red male is brightest on the throat, and is further distinguished by the pale bill. Bill shape and color also help to ID the bright-ochre female. Juvenile male initially resembles female, but gradually acquires more and more red coloration prior to spring migration. Frequently utters a staccato *chi-ti-duh*. A common migrant, from mid-Sept to late April. Individuals can be found countrywide in almost any habitat with some trees; to 8,200 ft (2,500 m). Range: Breeds in S US, winters south to N Bolivia.

uvenile male

female

Red-throated Ant-Tanager
Habia fuscicauda

male

female

7 in (18 cm). On dark male, note noticeably brighter and lighter throat. Likewise, the yellow throat is distinctive on the brownish female. Favors dense vegetation along streams and forest edges, where foraging groups produce a wren-like scolding chatter. The sweet, liquid song is also wren-like. Fairly common in Caribbean lowlands and foothills; to 3,000 ft (900 m). Range: SE Mexico to N Colombia.

Black-cheeked Ant-Tanager
Habia atrimaxillaris

female

7 in (18 cm). Male is similar to Red-throated Ant-Tanager (no geographic overlap), but with blackish sides of head. Female is similar to male, but with a more peach-colored throat. One of just three species found only in mainland CR. Small groups, often with mixed species flocks, move noisily through the lower levels of wet forest, occasionally coming to forest edges or into second growth. Combines harsh chatter with lilting, flute-like notes; easily mistaken for a wren. Relatively common in its limited range around the Golfo Dulce and Osa Peninsula. Range: CR.

Black-faced Grosbeak
Caryothraustes poliogaster

7 in (18 cm). The black face and throat set it apart from the somewhat similarly colored Prong-billed Barbet (p. 132), typically of higher elevations. Noisy groups of up to thirty or more birds travel through all levels of mature wet forest, forest edges, and adjacent gardens; often in mixed species flocks. Has a distinctive, sharp, buzzy call that it emits almost constantly while foraging. Common in Caribbean lowlands and foothills; to 3,000 ft (900 m). Range: SE Mexico to W Panama.

Black-thighed Grosbeak
Pheucticus tibialis

8 in (20 cm). The large bill and white wing spot set it apart from any other similarly plumaged species. Found in gardens and at forest edges. The call note is a sharp metallic *pwik*. The sweet song rambles up and down the scale with phrases that may end in a rapid trill or an emphatically repeated note. Fairly common in middle elevations and highlands, from Miravalles Volcano south, from 3,300 to 8,500 ft (1,000 to 2,600 m); occasionally descends to the Caribbean lowlands at mountain bases. Range: CR and W Panama.

Rose-breasted Grosbeak
Pheucticus ludovicianus

male

female

8 in (20 cm). The male's rose-red breast is definitive. The pale bill and white superciliary and wing bars distinguish the female and juvenile from other streaked-brown birds. Found in gardens and at forest edges. Gives a sharp, squeaky *eenk*. Fairly common migrant throughout the country, mostly from Oct to April; to 7,200 ft (2,200 m). Found in gardens and at forest edge. Range: Breeds in E Canada and US, winters south to Peru.

Blackbirds, Orioles (ICTERIDAE). Most species in this New World family have pointed, conical bills; while black is a predominant plumage color, many species sport splashes of bright color. Iris color can often be striking as well. A wide variety of habitats are occupied by the family, ranging from dry scrub to rainforest and from marshes to urban centers. These omnivorous birds feed on arthropods, small invertebrates, fruit, nectar, and seeds. In many species, the male is either larger or more colorful than the female. As a family, an array of nesting strategies are employed, from brood parasitism (the cowbirds) to monogamous pairs to polygamous colonial nesters. Typically, the female constructs the nest and incubates, while the male may contribute with defense and some provisioning of the young. Some species build open-cup nests, while others weave distinctive elongated pouches. World: 104, CR: 23

Red-winged Blackbird

Agelaius phoeniceus

male

female

9 in (23 cm). The male's scarlet-and-yellow epaulets are diagnostic. The bill shape and heavily streaked breast help differentiate the female from other brown birds. Inhabits marshes, wet fields, and edges of mangroves; often in flocks. Individuals frequently perch on roadside powerlines beside good habitat. Several sweet whistled notes are followed by a shrill, raspy *kreeEE!* Common in northern central Caribbean lowlands and from Palo Verde NP region south to Parrita; spreading southeast on both Pacific and Caribbean slopes. Range: NA to CR.

Melodious Blackbird
Dives dives

10 in (25 cm). The bill shape is the best feature for distinguishing it from any similar-sized, all-black bird. Pairs inhabit gardens in all climatic regions. Very vocal; whistles a loud, slurred *wheeur, wheeur, whit-wheeur*. First recorded in March 1987, now widespread and fairly common; to 7,200 ft (2,200 m). Range: E Mexico to CR.

Great-tailed Grackle
Quiscalus mexicanus

male

Male 17 in (43 cm); female 13 in (33 cm). The long tail and glaring yellow iris are striking features of both sexes. Commensal with humans almost anywhere there are even a few dwellings; walks with a swagger; hundreds of noisy individuals roost communally. Produces a variety of sounds ranging from shrill whistles to guttural grating noises. Common and widespread; to 6,600 ft (2,000 m). Range: S US to NW Peru.

female

Nicaraguan Grackle
Quiscalus nicaraguensis

male

female (left), juvenile (right)

Male 12 in (30 cm); female 10 in (25 cm). Very similar to much larger Great-tailed Grackle. Male is blacker, without bluish-and-green sheen, and tail is more deeply folded—especially noticeable in flight. Female is paler and grayer below than female Great-tailed Grackle. Inhabits riverbanks and borders of marshes. Uncommon in Caño Negro region, where greatly outnumbered by Great-tailed Grackle. Range: Nicaragua and CR.

Bronzed Cowbird
Molothrus aeneus

male

8 in (20 cm). The red iris sets apart the male from all other black birds, except the much larger and rather uncommon Giant Cowbird *M. oryzivorus* (not illustrated). Female is duller than male. Individuals or flocks of up to one hundred or more forage on the ground (in fields and along roadsides); often perches on utility wires. Females are brood parasites on a variety of passerine species. Fairly common and widespread; to 6,200 ft (1,900 m). Range: SW US to N Colombia.

Black-cowled Oriole
Icterus prosthemelas

adult (left), juvenile (right)

8 in (20 cm). Black-and-yellow adult is unique. Juvenile has yellow crown and back, black tail, and lacks wing bars. Favors forest edges, second growth, and gardens. Whistles a series of soft, sweet notes. Fairly common in Caribbean lowlands and foothills; to 4,300 ft (1,300 m). Recent scattered sightings (since 2003) in southern and central Pacific lowlands may indicate that range is expanding. Range: SE Mexico to W Panama.

Streak-backed Oriole
Icterus pustulatus

8 in (20 cm). One of two orange-headed orioles in dry forest habitat, it is distinguished by its streaked-back. (Spot-breasted Oriole *I. pectoralis* [not illustrated] has solid black back and black spotting on either side of black bib.) Pairs forage in dry forest canopy, at forest edges, and in gardens. An unaccomplished vocalist; gives a dry wren-like rattle. Fairly common in lowlands of northern Pacific; to 1,600 ft (500 m). Range: W Mexico to CR.

Baltimore Oriole
Icterus galbula

male

female

8 in (20 cm). Male's entirely black head and bright-orange underparts are diagnostic. Female and juvenile are duller orange and have white wing bars. Feeds on flower nectar and fruit at forest edges and in gardens. Call is a harsh, dry chatter. Common and widespread migrant, from early Sept to early May; to 7,200 ft (2,200 m). Range: Breeds in E NA, winters south to N SA.

Scarlet-rumped Cacique
Cacicus uropygialis

9 in (23 cm). Similar in coloration to male Passerini's and Cherrie's Tanagers (p. 212), but note pale-blue iris and pointed, pale bill. The scarlet rump is typically covered by the folded wings on perched birds. Two or more birds roam the middle and upper levels of mature wet forest, advanced second growth, and forest edges. Frequently gives loud, ringing whistles. Common in wet lowlands and foothills on both Caribbean and Pacific slopes; to 3,600 ft (1,100 m). Range: NE Honduras to SE Peru.

Chestnut-headed Oropendola
Psarocolius wagleri

Male 14 in (36 cm); female 11 in (28 cm). The pale bill, dull-chestnut head, and black wings set it apart from Montezuma Oropendola. Favors forests and forest edges, where individuals or flocks forage noisily at upper and middle levels of canopy. Uncommon in lowlands and middle elevations of Caribbean slope, rare in southern Pacific; to 5,600 ft (1,700 m). Range: SE Mexico to NW Ecuador.

oropendola nests

Montezuma Oropendola
Psarocolius montezuma

Male 20 in (51 cm); female 16 in (41 cm). The colorful bill and face, along with the bright-chestnut body, are diagnostic. Flocks forage at all levels in forests, forest edges, and adjacent gardens. Colonies nest in isolated trees, which are festooned with numerous large pouch nests. Abundant in Caribbean lowlands; colonies are more thinly distributed at middle elevations of the Caribbean slope, in the Central Valley, and along Pacific slopes of northern cordilleras; to 4,900 ft (1,500 m). Range: SE Mexico to central Panama.

Euphonias (FRINGILLIDAE). The composition of this widespread family has undergone some significant changes in recent decades with the inclusion of the euphonias (formerly considered tanagers) and the Hawaiian honeycreepers (formerly in their own family), thus making it somewhat difficult to generalize about the family. Nonetheless, in many species, the male is more colorful than the female. Bright plumage and/or pleasant vocalizations—canaries are also members of this family—result in the unfortunate fact that these birds are often sought as cage-birds, and Costa Rica is sadly no exception. World: 217, CR: 12

Yellow-crowned Euphonia

Euphonia luteicapilla

male

female

4 in (10 cm). The yellow extending to the rear of the crown separates the male from other male euphonias with dark throats and yellow on the forehead. The nondescript female is olive-green above and olive-yellow below. Favors forest edges and gardens. Calls repeatedly from high in trees with a shrill two- or three-note whistle: *sbee-sbee-sbee*. Common in wet lowlands and foothills; to 3,900 ft (1,200 m). Range: E Nicaragua to Panama.

Yellow-throated Euphonia
Euphonia hirundinacea

male

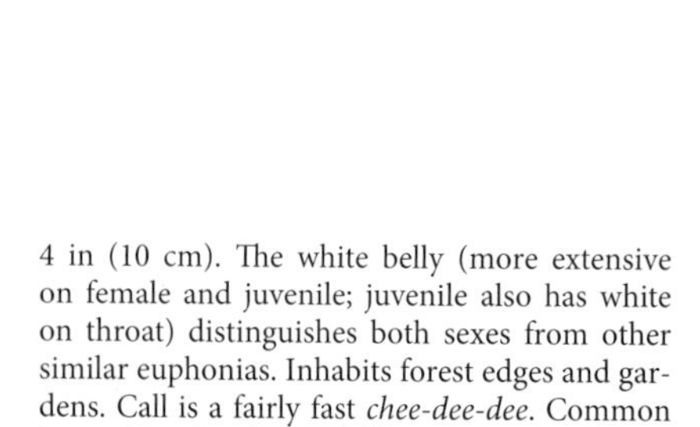

juvenile

4 in (10 cm). The white belly (more extensive on female and juvenile; juvenile also has white on throat) distinguishes both sexes from other similar euphonias. Inhabits forest edges and gardens. Call is a fairly fast *chee-dee-dee*. Common in northern Pacific foothills and in northern central Caribbean lowlands, uncommon in northern Pacific lowlands and Central Valley, rare in southern Pacific; to 4,600 ft (1,400 m). Range: E Mexico to W Panama.

Elegant Euphonia
Euphonia elegantissima

male

female

4 in (10 cm). Both sexes have distinctive powder-blue crown. Frequents forest edges and gardens. Makes a variety of not very musical sounds. Uncommon in middle elevations and highlands, from Tilarán Cordillera south; from 3,900 to 7,200 ft (1,200 to 2,200 m). Range: NW Mexico to W Panama.

Olive-backed Euphonia
Euphonia gouldi

male

female

4 in (10 cm). No other euphonia in its range has rufous on lower underparts. Mostly found at middle and lower levels of wet forest, forest edges, and gardens. The call is a somewhat slurred, shrill *shpree-ah-shpree*. Common in Caribbean lowlands and foothills; to 3,300 ft (1,000 m). Range: E Mexico to W Panama.

Golden-browed Chlorophonia
Chlorophonia callophrys

male

female

5 in (13 cm). The bright green upperparts and mostly yellow underparts distinguish both sexes. Often in forest canopy, though coming down to eye level at forest edges and in gardens. The soft, mournful, single-note whistle—lasting about one second—alerts one to the presence of these birds. Common from 3,300 ft (1,000 m) to timberline. Range: CR and W Panama.

Photo Credits

Obtaining great photos of birds—especially birds that are elusive, fast, or distant—is a challenging task. And, ideally, the photo is more than a great photo, it also shows key field marks. This book would not have been possible without the contribution of more than 50 talented photographers, whose national identities span the globe. We are especially grateful to a small group of Costa Rican photographers, without whose efforts we would never have obtained images of some of the rarer bird species.

The publisher wishes to make specific mention of Glenn Bartley and Jorge Chinchilla, who contributed so many beautiful photographs, and also Chris Jiménez, Raul Vega, Juan Carlos Vindas, and Doug Brown, whose stunning photographs have greatly improved this book. A heartfelt *thank you* also goes out to Lou Hegedus, who generously donated several photographs.

Please note that we have done our utmost to ascertain that none of the images is a composite image. Photoshop was used principally to adjust colors—and in other cases sparingly, to remove a twig here or erase a spot of mud there.

Front cover: Male Green-crowned Brilliant (*Heliodoxa jacula*), Gregory Basco
Spine: Male Green Honeycreeper (*Chlorophanes spiza*), Gregory Basco
Back cover, top to bottom: Keel-billed Toucan (*Ramphastos sulfuratus*), Orange-chinned Parakeet (*Brotogeris jugularis*), Brown-hooded Parrot (*Pyrilia haematotis*), Gregory Basco

p. ii: male Crowned Woodnymph (*Thalurania colombica*), Gregory Basco
p. vi: male Violet Sabrewing (*Campylopterus hemileucurus*), Gregory Basco
p. 244: male Long-tailed Manakin (*Chiroxiphia linearis*), Nick Hawkins
p. 248: male Red-winged Blackbirds (*Agelaius phoeniceus*), Jeffrey Muñoz

Species accounts:

Nick Athanas: p. 30 (top); p. 31 (bottom right); p. 32 (Royal Tern, bottom); p. 35 (bottom); p. 37 (bottom); p. 38 (top); p. 106 (top); p. 111 (top); p. 151; p. 157 (top); p. 169 (bottom); p. 189; p. 194 (top); p. 201 (bottom); **Glenn Bartley**: p. 18 (top); p. 19 (top); p. 20 (top & bottom left); p. 21 (top); p. 23 (bottom); p. 24 (top and bottom); p. 25 (top, bottom, & middle); p. 26 (bottom); p. 27 (bottom left); p. 28 (top & bottom); p. 29 (top & bottom); p. 33 (bottom); p. 34 (top); p. 35 (top); p. 36 (top & bottom); p. 37 (top); p. 38 (bottom); p. 40 (bottom); p. 41 (top); p. 42 (top & bottom); p. 43 (top);

p. 44 (bottom); p. 46 (bottom); p. 47 (top, middle, & bottom); p. 48; p. 49 (bottom); p. 50 (top, middle, & bottom); p. 52 (top); p. 53 (top, middle, & bottom); p. 54 (all photos); p. 57 (top & middle); p. 58 (top); p. 59 (middle); p. 61 (top left); p. 65 (bottom left); p. 72 (all photos); p. 73 (bottom left); p. 74 (all photos); p. 75 (top); p. 76 (top & bottom); p. 78; p. 80; p. 81 (top); p. 82 (bottom); p. 86 (bottom); p. 87 (all photos); p. 88 (Green Thorntail, bottom); p. 88 (Black-crested Coquette, top & bottom); p. 90 (top); p. 91 (top left & top right); p. 92 (White-throated Mountain-gem, bottom); p. 94 (bottom); p. 95 (middle); p. 96 (top left, middle, & bottom); p. 98 (top); p. 102 (top); p. 103 (top); p. 105 (top); p. 107; p. 109 (bottom); p. 113 (top & bottom); p. 116 (bottom); p. 117 (left); p. 118 (top left & top right); 124 (bottom); p. 125 (bottom); p. 126 (all photos); p. 128 (bottom); p. 129 (top); p. 131 (top); p. 133; p. 135 (top & bottom); p. 137 (top right); p. 139 (bottom left); p. 140 (top left); p. 142 (bottom); p. 145 (top); p. 149 (bottom); p. 150 (top & middle); p. 154 (top); p. 155 (top); p. 156 (top); p. 162 (top & bottom); p. 163 (bottom); p. 165 (top); p. 166 (top & middle); p. 168 (top); p. 170 (top); p. 179 (bottom); p. 185 (top); p. 193 (top); p. 194 (bottom); p. 196 (bottom); p. 198 (top); p. 200 (top & bottom); p. 202 (middle); p. 204 (Yellow Warbler [Northern], top & bottom); p. 205 (top); p. 206 (bottom); p. 207 (middle & bottom); p. 212 (Passerini's Tanager, top); p. 212 (Cherrie's Tanager, top & bottom); p. 214 (top & bottom); p. 216 (middle & bottom); p. 217 (top & bottom); p. 218 (bottom left & bottom right); p. 219 (Green Honeycreeper, top); p. 222 (bottom); p. 223 (bottom); p. 226 (top & middle); p. 234 (top & bottom); p. 235 (top & bottom); p. 236 (middle & bottom); p. 238 (bottom); p. 239 (top); p. 240 (bottom); p. 241 (bottom); p. 243 (Golden-browed Chlorophonia, top & bottom); **Gregory Basco**: p. 19 (bottom); p. 20 (bottom right); p. 218 (middle); **Cindy Beckman**: p. 210 (bottom); **David B. Bernstein**: p. 62 (top); p. 192 (middle); **Doug Brown**: p. 31 (top left); p. 32 (Royal Tern, top); p. 32 (Sandwich Tern, top); p. 33 (top); p. 39 (bottom); p. 41 (bottom); p. 44 (top); p. 60 (top & bottom); p. 67 (top right); p. 85 (top & middle); p. 95 (top); p. 97 (top); p. 103 (bottom); p. 110 (top); p. 112 (top & bottom); p. 122 (top & bottom); p. 137 (top left, bottom left, & bottom right); p. 208 (bottom); p. 213 (top); p. 224 (top); p. 229 (bottom); p. 231 (top); p. 242 (Yellow-throated Euphonia, top); **Mauricio Calderón**: p. 178; **Les Catchick**: p. 236 (top); **Phoo Chan**: p. 159 (middle); **Jorge Chinchilla**: p. 22 (middle & bottom); p. 46 (top); p. 59 (bottom); p. 69 (bottom); p. 73 (bottom right); p. 79 (top); p. 81 (bottom); p. 82 (middle); p. 84 (top & bottom); p. 88 (Green Thorntail, top); p. 89 (all photos); p. 98 (bottom left & bottom right); p. 100 (bottom); p. 102 (bottom); p. 108 (bottom); p. 109 (top); p. 110 (middle); p. 114; p. 115 (bottom); p. 119 (top left, top right, & bottom right); p. 120 (top left, top right, & bottom left); p. 121 (top & bottom left); p. 124 (top); p. 132; p. 136 (top & bottom); p. 138 (top left & bottom right); p. 140 (bottom left & bottom right); p. 143 (top & bottom); p. 144; p. 145 (bottom); p. 146 (top & bottom); p. 149 (top); p. 150 (bottom); p. 153 (bottom); p. 154 (bottom); p. 157 (middle & bottom); p. 158 (bottom); p. 164 (top); p. 165 (bottom); p. 167 (bottom); p. 169 (top); p. 172 (top & bottom); p. 173 (middle & bottom); p. 174 (top); p. 175 (top); p. 176 (Red-capped Manakin, top); p. 176 (White-collared Manakin, bottom); p. 182; p. 185 (bottom); p. 187 (bottom); p. 197 (top & bottom); p. 198 (bottom); p. 199 (bottom); p. 202 (bottom); p. 203 (top); p. 205 (middle); p. 207 (top); p. 208 (top); p. 211 (bottom); p. 215 (top & bottom); p. 216 (top); p. 218 (top); p. 219 (Green Honeycreeper, bottom; Slaty Flowerpiercer, top); p. 220 (Blue-black Grassquit, bottom; Thick-billed Seed-Finch, bottom); p. 222 (top & middle); p. 223 (top & middle); p. 225 (top); p. 226 (bottom); p. 227 (top & bottom); p. 228 (top); p. 230 (all photos); p. 232 (top & middle); p. 233 (top & bottom); p. 239 (middle); p. 243 (Olive-backed Euphonia, top); **Brett Cole**: p. 61 (middle); **David Comings**: p. 118 (bottom right); **Mike Danzenbaker**: p. 18 (bottom); p. 105 (bottom); p. 148 (top); p. 149 (middle); p. 152; p. 232 (bottom); **Paul Ellis**: p. 92 (Magenta-throated Woodstar, bottom); © **Kip Evans Photography**: p. 67 (top left); **Fritz Fucik**: p. 93 (top); **Richard Garrigues**: p. 27 (top); p. 31 (bottom left); p. 39 (top); p. 43 (bottom); p. 57 (bottom); p. 58 (bottom); p. 63 (top); p. 67 (bottom right); p. 77 (bottom); p. 101 (bottom); p. 118 (bottom left); p. 123 (bottom); p. 129 (bottom); p. 138 (bottom left); p. 140 (top right); p. 147 (Fasciated Antshrike, top); p. 147 (Barred Antshrike, bottom); p. 161 (all photos); p. 167 (top); p. 169 (middle); p. 187 (top); p. 188; p. 203 (bottom); p. 212 (Passerini's Tanager, bottom); p. 219 (Slaty Flowerpiercer, bottom); p. 237 (top); p. 240 (middle); p. 242 (Yellow-throated Euphonia, bottom); **Steven Garvie**: p. 21 (bottom); **Bob Gress**: p. 67 (bottom left); p. 201 (top); **Alan Harper**: p. 155 (bottom); **Wang Hc**: p. 177 (top); **Lou Hegedus**: p. 27 (bottom right); p. 56 (top); p. 111 (bottom); p. 125 (top); p. 134 (bottom); p. 138 (top right); **Adrian Hepworth**: p. 30 (bottom right); p. 41 (middle); p. 45 (bottom); p. 91 (middle); **Yehudi Hernández**: p. 121 (bottom right); p. 210 (top); **Oscar Herrera**: p. 99 (top); **Chris Jiménez**: p. 51; p. 62 (bottom left & bottom right); p. 64 (top & middle); p. 65 (top & middle); p. 66 (top & bottom); p. 68 (top); p. 71 (top); p. 73 (top); p. 75 (bottom); p. 85 (bottom); p. 86 (top); p. 91 (bottom); p. 92 (Magenta-throated Woodstar, top); p. 93 (bottom right); p. 94 (middle); p. 95 (bottom); p. 99 (bottom); p. 108 (top); p. 110 (bottom); p. 127 (top & bottom); p. 130 (top & bottom); p. 157 (middle & bottom); p. 166 (bottom); p. 176 (White-collared Manakin, top); **Tom Johnson**: p. 209; **Nick Kontonicolas**: p. 192 (bottom); **Brian Kushner**: p. 40 (top); **Paul Kusmin**: p. 90 (bottom); **Eduardo Libby**: p. 180 (top); **Lee Marcus**: p. 69 (top); **Chris Morgan**: p. 177 (bottom); **Alex Navarro**: p. 159 (top); **Richard Orr**: p. 22 (top); **Judd Patterson**: p. 148 (bottom); **David Pereksta**: p. 199 (top); **Brian Pollock**: p. 153 (top); **Michael Retter**: p. 180 (bottom); **Roger Rodgriquez**: p. 158 (top); **Joel N. Rosenthal**: p. 116 (top); **Yamil Sáenz**: p. 174 (bottom); **Ed Schneider**: p. 159 (bottom); **Marek Stefunko**: p. 55 (bottom); **Stan Tekiela**: p. 119 (bottom left); **Roy Toft**: p. 128 (top); **Noel Ureña**: p. 45 (top); p. 61 (bottom); p. 63 (bottom left & bottom right); p. 96 (top right); p. 97 (bottom); p. 100 (top); p. 168 (bottom); p. 179 (top); p. 193 (bottom); p. 206 (top); p. 221 (middle); p. 228 (bottom); **Alex Vargas**: p. 34 (bottom); p. 70 (all photos); p. 73 (middle); p. 79 (bottom); p. 104 (top); p. 164 (middle & bottom); p. 192 (top); p. 204 (Yellow Warbler [Mangrove], bottom); **Luis Vargas Durán**: p. 31 (middle left); p. 31 (top right); p. 32 (Sandwich Tern, bottom); p. 147 (Fasciated Antshrike, bottom); p. 173 (top); p. 237 (middle); **Raul Vega**: p. 26 (top); p. 61 (top right); p. 71 (bottom); p. 82 (top); p. 101 (top); p. 106 (bottom); p. 123 (top); p. 170 (bottom); p. 175 (bottom); p. 181 (bottom); p. 183 (top & bottom); p. 184 (top); p. 186 (top & bottom left); p. 190 (top & bottom); p. 191 (bottom); p. 202 (top); p. 204 (Yellow Warbler [Mangrove], top); p. 205 (bottom); p. 211 (top & middle); p. 220 (Blue-black Grassquit, top); p. 224 (bottom); p. 225 (bottom); p. 228 (middle); p. 229 (top); p. 237 (bottom); p. 238 (top); p. 239 (bottom); p. 242 (Elegant Euphonia, top & bottom); p. 243 (Olive-backed Euphonia, bottom); **Alexander Viduetsky**: p. 30 (bottom left); **Juan Carlos Vindas**: p. 23 (top); p. 49 (top); p. 52 (bottom); p. 55 (top); p. 59 (top); p. 64 (bottom); p. 65 (bottom right); p. 68 (bottom); p. 77 (top); p. 83; p. 92 (White-throated Mountain-gem, top); p. 93 (middle & bottom left); p. 94 (top); p. 104 (middle & bottom); p. 115 (top); p. 117 (right); p. 120 (bottom right); p. 131 (bottom); p. 134 (top); p. 139 (top left, top right, & bottom right); p. 141; p. 142 (top); p. 147 (Barred Antshrike, top); p. 156 (bottom); p. 160 (top & bottom); p. 163 (top); p. 171; p. 179 (middle); p. 181 (top & middle); p. 184 (bottom); p. 186 (bottom right); 191 (top); p. 195 (top & bottom); p. 196 (top); p. 213 (bottom); p. 220 (Thick-billed Seed-Finch, top); p. 221 (top & bottom); p. 231 (bottom left & bottom right); p. 240 (top); p. 241 (top); **Ray Wilson**: p. 176 (Red-capped Manakin, bottom); **Phil Yates**: p. 56 (bottom).

Species Index

About the Author

Richard Garrigues has been birding since the age of sixteen, when a close encounter with a Black-and-white Warbler walking up a tree trunk just a few feet away from him in suburban New Jersey made a lasting impression. Since 1981, he has lived in Costa Rica, where for nearly thirty years he has been leading birding and natural history tours.